D0616438

# How to Change Minds
ABOUT
# Our Changing Climate

THE EXPERIMENT

BECAUSE EVERY BOOK IS A TEST OF NEW IDEAS

# How to Change Minds
## ABOUT
# Our Changing Climate

LET SCIENCE DO THE TALKING THE NEXT TIME
SOMEONE TRIES TO TELL YOU . . .

The Climate Isn't
Changing

Climate Change Is
Natural, Not Man-Made

Global Warming
Is Actually a
Good Thing

. . . AND OTHER ARGUMENTS IT'S TIME TO END FOR GOOD

**SETH B. DARLING AND DOUGLAS L. SISTERSON**

**ILLUSTRATIONS BY SARAH M. SISTERSON**

**THE EXPERIMENT**
NEW YORK

How to Change Minds About Our Changing Climate: *Let Science Do the Talking the Next Time Someone Tries to Tell You . . . The Climate Isn't Changing; Global Warming Is Actually a Good Thing; Climate Change Is Natural, Not Man-Made . . . and Other Arguments It's Time to End for Good*
Copyright © Seth B. Darling and Douglas L. Sisterson, 2014
Illustrations copyright © Sarah M. Sisterson, 2014

The Experiment, LLC
220 East 23rd Street • Suite 301
New York, NY 10010-4674
www.theexperimentpublishing.com

The Experiment's books are available at special discounts when purchased in bulk for premiums and sales promotions as well as for fund-raising or educational use. For details, contact us at info@theexperimentpublishing.com.

Many of the designations used by manufacturers and sellers to distinguish their products are claimed as trademarks. Where those designations appear in this book and The Experiment was aware of a trademark claim, the designations have been capitalized.

Library of Congress Cataloging-in-Publication Data

Darling, Seth B.
  How to change minds about our changing climate : let science do the talking the next time someone tries to tell you ... : the climate isn't changing : global warming is actually a good thing : climate change is natural, not man-made : ... and other arguments it's time to end for good / Seth B. Darling and Douglas L. Sisterson ; illustrations by Sarah M. Sisterson.
     pages cm
  Includes bibliographical references and index.
  ISBN 978-1-61519-223-6 (pbk.) -- ISBN 978-1-61519-224-3 (ebook)
1. Climatic changes--Social aspects. 2. Global warming--Social aspects. 3. Environmental education. I. Sisterson, Douglas L. II. Title.
  QC981.8.G56D37 2014
  363.738'74--dc23
                          2014014512

ISBN 978-1-61519-223-6
Ebook ISBN 978-1-61519-224-3

Cover design by Orlando Adiao
*Text design by Pauline Neuwirth, Neuwirth & Associates, Inc.*

Manufactured in the United States of America
Distributed by Workman Publishing Company, Inc.
Distributed simultaneously in Canada by Thomas Allen and Son Ltd.

First printing July 2014
10 9 8 7 6 5 4 3 2 1

*To our children, Isaac Darling and*
*Nathaniel, Sarah, and Rachel Sisterson*

# Contents

# PART 5: There's Nothing We Can Do About It

# Introduction

**IS GLOBAL WARMING** just the result of natural cycles? Or of cosmic rays bombarding the Earth? Climate change skeptics say yes to questions such as these routinely and loudly. Have you heard the claim that scientists are split on whether humans are affecting the climate? Have you ever wondered if renewable energy is too expensive to replace fossil fuels? Skeptic-inspired misconceptions about climate change such as these are everywhere: in recent statements from leading politicians, in public opinion polls, and in "balanced" news coverage that often goes out of its way to give equal weight to science and skepticism. We are regularly confronted with arguments that deny climate change is happening or is a problem; these claims come from many directions, including news reports on TV and radio, newspapers and blogs, and even sometimes in direct conversations with climate-change skeptics.

As a result of this divergence of messages, the general public is understandably confused. Though public consensus is slowly building that climate change is happening and is caused by human activity (see Chapter 1), most non-scientists (and even most scientists) are not equipped to know what's true and what's false when faced with the assertions of adamant climate-change skeptics.

We are both researchers at a US Department of Energy (DOE) research laboratory,* and we both frequently deliver public lectures

---

*This is where the lawyers make us say this: the views expressed in this book are those of the employees and not those of Argonne, UChicago–Argonne, the University of Chicago, or DOE.

on energy and climate to all sorts of different audiences. Every once in a while an adamant skeptic is in the crowd. On one occasion, I (Seth) was confronted by a skeptic who raised a myth that—at the time—was new to me, and as a result, I was not prepared to offer any evidence disputing his claim. His argument was based on what sounded like reasonable scientific rationale involving saturation of the effect of carbon dioxide ($CO_2$) on greenhouse warming (see Chapter 12 for an explanation of this misconception and why it is wrong). Being ill-equipped to handle this specific myth, as it was novel to me, I couldn't crisply reveal the errors in the skeptic's argument. That experience haunts me to this day, because I suspect that there were people in that audience who left believing that his case was credible. Experiences like this have inspired us not only to educate ourselves on the full spectrum of skeptic claims but also to write this book. In what follows, we distill years of research into a concise compilation of scientific explanations refuting climate-change skepticism. Because outreach through public lectures and the like—though something we enjoy thoroughly—only reaches one classroom or lecture hall at a time, our hope is that this book will help educate an army of voices, each equipped to change minds one conversation at a time.

In the following chapters, we aim to provide a response to the plethora of skeptic misconceptions that inundate the media and blogosphere. Using clear and accessible explanations of what we do and don't know about the science, we hope to equip readers with the tools to distinguish fact from fiction, to see through the smoke and mirrors, and to understand what needs to be done to address climate change and why. You don't need a degree in science to understand the basic principles of climate change, but you do need to have some facts straight—facts that we're confident you'll have at hand after reading this book.

There are indeed things we still don't know about our planet's climate and our effect on it, but the basic tenets attacked by skeptics are generally those that the scientific community has established with mountains of evidence. Nearly all skeptic arguments are based on a common error: cherry-picking pieces of data without seeing the big picture. It's what lies behind the claim that the glaciers are growing, or, say, the assertion that the planet isn't getting warmer. The error of cherry-picking data carries through the book, as we devote one chapter to each of the main skeptic myths and succinctly bust those myths with accounts of the facts.

We have assembled a comprehensive list of climate-skeptic myths, some put forward by famous skeptics and others that we've run across in our interactions with the public or in online discussion forums. We've grouped them into sets with shared themes, ranging from the confusion of climate with weather, through claims that global warming isn't such a bad thing and that carbon taxes will kill jobs and hurt the poor, to the belief that reports about climate change are all parts of a conspiracy. New myths are sure to crop up in the future, but readers will learn from this book that skeptic myths tend to suffer from common faults. Data are cherry-picked. Timescales are confused. Cause and effect are muddied. Recognizing these common errors will allow readers to see through whatever myths emerge.

There are other lessons embedded here, too—about scientific consensus, interdependent phenomena, and the importance of understanding the big picture—that we believe apply well beyond climate change, to the practice of science as a whole. There is an important role for skepticism in science, but skeptics' arguments regarding climate change are usually governed more by money and politics than by the rules of scientific reasoning and consensus.

Stepping back, regardless of political perspective, regardless of whether one believes it is happening, we see that our planet's climate is indeed changing, and that we are indeed to blame. Effects of climate disruption are already apparent in everything from rising seas to more extreme weather to the fact that native plants once flourishing in our backyards are dying as the local climate zones shift. The responsibility for maintaining the Earth's climate lies with each of us; while power plants may be some of the largest emitters of greenhouse gases, we are all consumers, in one way or another, of the power they produce. This isn't about pointing fingers. It's about identifying and understanding the problem and, more important, taking action to do something about it. Failure to do so puts our children's future—and that of all subsequent generations—in peril. Alarm at such a prospect motivates our own research as well as our outreach efforts, including authoring the book in your hands.

# Prologue
## THE BASICS

**THE EARTH GETS** nearly all its warmth from the sun's radiant energy. The side of the planet facing the sun heats up, and then, as the Earth rotates and that side turns away from the sun and is shrouded in darkness, the planet returns much of that energy back to space as heat. If the Earth's atmosphere were completely transparent, allowing the Earth to return all the sun's energy back to space at night without any heat getting trapped in the atmosphere, the average surface temperature of the planet would be about zero degrees Fahrenheit—that's right, thirty-two degrees below the freezing point of water! Fortunately for us, the atmosphere contains gases that keep some of the heat in. These are the greenhouse gases, and their warming influence is called the greenhouse effect.

Our planet's atmosphere is composed mostly of nitrogen and oxygen; only about 0.05 percent (by mass) of the Earth's atmosphere contains greenhouse gases. There are about a dozen different greenhouse gases that trap heat by absorbing thermal infrared radiation that would otherwise be returned to space by our planet at nighttime.* This seemingly negligible concentration of greenhouse gases has a gigantic impact on the surface temperature of

---

*The primary greenhouse gases on Earth are carbon dioxide, methane, nitrous oxide, ozone, and various chlorofluorocarbons (CFCs). Water vapor, which makes up about 0.4 percent of the atmosphere, is also a significant greenhouse gas, but the overall climate effect of water vapor is complex, involving both warming and cooling influences. Moreover, the amount of water vapor in the atmosphere depends significantly on the amount of the other greenhouse gases. Check out Chapter 11 for more on this.

our planet. The combined effect of this tiny amount of greenhouse gas results in the average temperature of the Earth's surface being about sixty degrees Fahrenheit (sixteen degrees Celsius).* While we'd sure save some money on air-conditioning without the greenhouse effect, we'd also all be dead. Not a fair trade. On the flip side, relatively small increases in the concentration of greenhouse gases in the atmosphere can push the temperatures up even higher.

Before we get into changes to the Earth's climate, we first have to understand what "climate" really means. The climate where you live is called regional climate; it is simply the thirty-year average of weather in one place. Global climate is the average of all the regional climates of the world. The World Meteorological Organization (WMO), an agency of the United Nations, is an intergovernmental organization charged with being the authoritative voice on issues related to meteorology, hydrology, and related sciences. The WMO mandates each member nation to compute thirty-year averages of meteorological quantities at least every thirty years (1931–1960, 1961–1990, 1991–2020, and so on). These averages are called "climate normals." Meteorologists and climatologists regularly use normals for putting recent climate conditions into a historical context. Normals were not designed to be metrics of climate change. In fact, when the widespread practice of computing normals began in the 1930s, the generally accepted notion of the climate was that the underlying long-term averages were constant. (More on this at the end of the chapter.)

So why thirty years? It seems like an arbitrary length of time. Surely one year is not a good representation of the climate in a region, since it could be an unusually hot, cold, dry, or wet year for any of a number of reasons (many of which we'll discuss in later chapters). Same goes for two or, say, five years. A one-hundred-year period would certainly give a nice average, but folks would

---

*Clouds and other factors also play important roles in regulating the planet's temperature.

probably accuse the WMO of being lazy if they only got around to assembling climate data once a century. Thirty it is.

Why did they start with 1930? Well, by 1930, there were simply enough weather stations around the world that computing thirty-year averages was feasible. These thirty-year averaged data give us a reference for what we might expect of our climate both now and in the future—as long as the amount of greenhouse gases in our atmosphere and the sun's radiant energy stay the same.

Data and measurements like those taken at weather stations around the world are a bedrock of climate science. We have only limited data prior to 1930 and, of course, only educated guesses for what will happen hundreds to thousands of years from now. So, in terms of deciphering Earth's climate history, we are currently in something of an Oreo era: We are making good measurements now (data are the cream filling), but these great data are sandwiched between large periods of uncertainty. (Apologies to those who prefer the cookies to the cream.)

To understand what climate is, it can be helpful to understand what it isn't. Perhaps the most common misconception about climate is that it is what's happening outside your window right now. That's weather: a day-to-day reflection of our climate where we live. To establish global climate normals, we basically average local weather for thirty years and average all those averages over the entire planet to define the Earth's climate at the global scale. Weather is defined by temperature, pressure, wind speed, wind direction, rainfall or snowfall amount, cloudiness, etc. To describe the regional climate of a place, people often tell what the temperatures are like over the seasons, how windy it tends to be, and how much rain or snow falls on average. So climate and weather are two different things, but because one is essentially the long-term average of the other, they are not independent of each other. (Chapter 6 deals with this distinction in more detail.)

All right, so now that we're all on the same page about what climate is (and isn't), let's focus on climate change. What does it mean when we say that the climate is changing? First, you might be wondering, is "global warming" just another name for the same thing? Indeed, these terms are often used interchangeably in the vernacular. Technically, though, "global warming" refers to surface-temperature increases, whereas "climate change" includes global warming as well as other long-term climatic effects. While we're on the subject of semantics, it's important to keep in mind that global warming is a two-word term, one of which is "global." Locally, climatic effects can vary from global trends. We should not get locked into the notion that global warming, at least in the short run, results in a rise of temperature in every climatic region.

Climate change may be a hot topic today (no pun intended), but it is not something new. Our planet's climate has undergone some pretty big changes before. As with so many things, we can learn a lot about the future by looking into the past. Let's spend a moment considering the history of our planet's climate. (We're talking about a 4.5-billion-year-long history, so a moment isn't really fair, but we'll see what we can do.) But wait, you say. If we only started to collect rigorous weather (and therefore climate) data on a world scale in the 1930s, how do we know what the climate and weather of our planet were like billions of years ago?

Since the beginning of Earth's history, climate has varied on many different timescales. Over millions of years, it has swung between very warm conditions, with annual mean temperatures above fifty degrees Fahrenheit (ten degrees Celsius) in polar regions, and glacial climates in which the ice sheets covered the majority of the mid-latitude continents. It has even been suggested that, in some past cold periods, the whole surface of the Earth was covered by ice. But over shorter timescales, lower-amplitude fluctuations occur, with no year being exactly the same as a previous one.

Extracting climate data from millions to billions of years ago involves tremendous detective work, but looking at relatively more recent deep history, we've got some surprisingly good data in the form of ice cores. Every year, snow falls on the ice caps and glaciers around the world, and these annual layers build up on top of one another like the rings of a tree, preserving information about the local temperature, atmospheric composition, and other useful nuggets of data. By digging deeper down into the ice sheet, you get information from further into the past. Depending on the ice core, scientists can collect data dating back anywhere from just a few years to as long as 800,000 years ago!*

Scientists have used ice-core samples to obtain some very basic data about local surface temperature and the composition of the lower atmosphere—with levels of greenhouse gases such as carbon dioxide being of particular interest in the context of climate change. These data show cyclical behavior—which represents glacial cycles—over the course of hundreds of thousands of years. Temperature and carbon dioxide track each other to an amazing degree throughout the ice-core record. The approximate level of carbon dioxide over that time period is 230 parts per million, with peaks at 250–300 parts per million about every 100,000 years.

The most recent increase, however, is dramatic. Carbon dioxide levels started to climb noticeably beginning in the late eighteenth century—the dawn of the Industrial Revolution—and have continued their upward march since then, to about 400 parts per million today. In other words, our current atmosphere has more carbon dioxide in it than at any time in the entire history captured in the ice cores. In fact, we know from various carbon dioxide proxy measurements, such as isotope ratios in marine sediments and the number of stomata on fossil plant leaves, that this level of carbon dioxide has not been seen on Earth since the Pliocene Epoch about three million years ago. What was the Earth's climate like back then? Let's just say it didn't look much like it does now, and it definitely was not an Earth amenable to modern society. For example, sea level was about twenty-five meters (eighty-two feet) higher than

---

*The current record holder is the European Project for Ice Coring in Antarctica (EPICA) core, which captures information about eight previous glacial cycles.

it is today! (Check out Chapter 3 for some of the consequences of a hotter planet.)

The fact that atmospheric carbon dioxide levels have climbed in concert with industrial activity sure seems too conspicuous to be pure coincidence, but to understand how we got into the situation

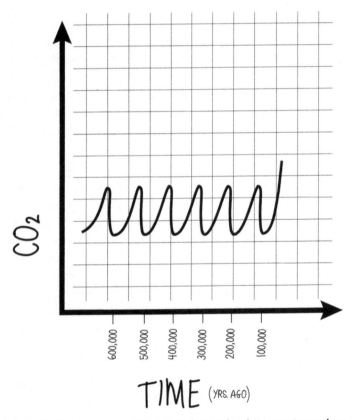

Schematic of typical ice core data showing large swings in temperature and carbon dioxide level—tracking each other—during a series of glacial cycles

we're in—to understand climate change—we'll first have to talk a bit about energy. We use energy at a breathtaking rate. The standard unit we use to describe the rate of energy use is the watt, which is a unit of energy (joule) per unit of time (second). In other words, a watt is not a unit of energy per se; rather, it's a unit describing how rapidly energy is used. You can think of it like miles vs. miles-per-hour. Worldwide, we consume energy today at an average rate of about 18 trillion joules every second—that is, 18 trillion watts, or 18 terawatts (TW). Using 18 TW is like running 18 billion microwave ovens all the time. You could make enough popcorn to stack up to, well, something absurdly far away.

The thing is, not only do we use massive amounts of energy, but we also keep increasing the rate of our energy use every year. In 1990, the rate was more like 12 TW—we needed a steady supply of about 12 TW of power throughout that year. Back in 1950? A little over 3 TW. This trend is projected to continue for many decades to come. The growth in demand is driven primarily by population growth and the developing world because of, well, development. In other words, assuming the developing world doesn't undergo a catastrophic change of course, we will need a lot more energy in the future. To give you a sense of where we're headed, the projected global demand in 2050 is 30 TW—that's getting close to twice the rate of energy use today. So where does our energy come from today, and, more important, from where are we going to get 30 TW of it by 2050?

Energy is a commodity. For the most part, the market doesn't really care where the energy comes from—it cares what the energy costs. This means that we opt for energy sources that are plentiful and relatively easy to use—and, therefore, cheap. Fortunately, our planet has vast resources of energy-rich materials available to us: coal, petroleum, and natural gas. These sources are collectively called fossil fuels because they are formed by the decomposition of organisms that died millions of years ago and were buried under heavy sediments. They represent about 83 percent of today's overall global energy supply. The next-largest sources of energy are hydropower and nuclear energy, representing 8 percent and 6 percent of the mix, respectively. We'll save you the arithmetic and tell you that the rest of the sources only provide tiny contributions to our energy mix.

Fossil fuels, like the organisms from which they came, are composed largely of carbon and contain energy that has been locked up for eons. Burning fossil fuel releases this energy, which we use to get our cars to the supermarket or to generate electricity to watch *Law & Order* reruns. It also releases lots of other stuff—most of it unpleasant in one way or another. Air pollutants like nitrogen oxides, sulfur dioxide, heavy metals, radioactive elements, and particulates damage ecosystems, infrastructure, and human health on a colossal scale, but they aren't the subject we're focusing on here. Coal ash, thermal pollution, and hydraulic fracturing threaten our fresh water systems, but again that's fodder for some other book. Disasters like oil spills, mine collapses, and gas explosions take human lives and inflict billions of dollars of costs on all of us, but they aren't the subject here, either. Let's get to the point. When you burn a carbon-containing material like coal, a major product of the reaction is . . . you guessed it: carbon dioxide.

We're talking about truly impressive amounts of carbon dioxide here, something in the ballpark of thirty billion tons per year. This carbon dioxide ends up in lots of places. The oceans absorb huge amounts of carbon dioxide, making carbonic acid and thereby increasing their acidity. Acidic oceans are unfriendly to coral reefs, shellfish, and other organisms—that's a topic for some other book, too. Soils also suck up some of the carbon dioxide, as do plants. The part that's of concern for our conversation, though, is the carbon dioxide that stays in the atmosphere. You'll recall that carbon dioxide is a greenhouse gas. More greenhouse gas in the atmosphere means a stronger greenhouse effect (see Chapter 10).

OK, so fossil fuels are cheap but have some serious side effects. Where are we supposed to turn to find 30 TW of energy without going all drill-baby-drill? One thing to get straight up front is that it takes a long time to shift energy sources on a global scale, and in all likelihood we are still going to be using fossil fuels in 2050. But that's not to say we shouldn't look at the other options. What are they?

Nuclear fission, which has a low carbon footprint, could be scaled up big-time, though that would require addressing concerns about waste,* weapons proliferation, disasters, and terrorist plots—and it wouldn't be cheap, either. Nuclear *fusion* technology is, as the joke goes, thirty years away—always. (Trust us, that's funny to energy nerds, though perhaps not so much to fusion research-ers.) Hydropower could be expanded a bit, but it is largely tapped out already. That leaves wind, solar,† biomass, geothermal, and the oceans (see Chapter 15). There are practical limitations to how much energy we could feasibly extract from each of these. We'll return to this question at the end of the book, but we'll toss you a bone so you don't get too glum after all this bad news: We have alternative options, and they can work.

Now that we've got our terminology straight and a general pic-ture of climate and its connection with energy use, allow us to introduce you to Brad.‡ You may know a Brad (or two or three). He's that neighbor/brother-in-law/coworker/politician/TV personality/ blogger who thinks this whole climate change thing is a bunch of malarkey, and he's got arguments—and even some data—to back up his claims. He is a climate-change skeptic, and if you're a mem-ber of the majority of folks who accept that climate change is hap-pening and that it's caused by human activity (see Chapter 1), he can be an exasperating thorn in your side. Brad may have ideological

---

*Next-generation nuclear fission reactors could alleviate this issue somewhat by burning the really nasty stuff, though there would still be radioactive waste to deal with.

†Solar energy researchers have a fusion joke of their own: They believe that nuclear fusion is the answer to our energy needs, but they like the reactor being 93,000,000 miles away. Yeah, we can hear the groans.

‡Apologies if you or someone you love is named Brad!

motives, or maybe he's just fallen victim to the very same faulty arguments that he's parroting to you.

Odds are that you aren't a climatologist or a research meteorologist, so maybe Brad's latest tirade sends you hunting for some answers. (Some of his ideas might be so out-of-left-field . . . or more likely right field in this case . . . that even a trained climate researcher might not have a well-formed response immediately ready.) This handbook provides you with those answers. We've scoured the skeptic arguments, or myths, as we'll frequently refer to them here, and supplied reliable, easy-to-follow responses to them based on the facts as researchers understand them today. This is a myth-busting book.

Let's start with our first myth right here, right now.

Skepticism is supposed to be a good thing in science, right? So isn't Brad the Galileo of our time, pulling the curtain back on a huge scientific blunder? We didn't make that up . . . this very claim has been made repeatedly by climate skeptics, and we'll get to it in a moment. But first there are a few things to get straight. There are many things we don't know about climate change, but the premise that the climate is changing is the only explanation available that is consistent with the data we have to date. Science works by consensus, but it is rare that any scientific argument has 100 percent consensus. The process that scientists use to get consensus is referred to as the scientific method. The heart of the matter is that it is not sufficient for a scientist to *believe* he or she is right; nor is it sufficient for a scientist to *know* he or she is right. Scientists must *convince* their colleagues that they are right.

Galileo, Brad? Really? Contrary to the common perception today of Galileo as a one-man outdated-tradition-busting machine, Galileo was a proponent of a conclusion many of his contemporaries actually agreed with: The Earth orbits the sun rather than the other way around. A consensus for this so-called heliocentric view was building at the time. A key point is that this conclusion was based on evidence, whereas those of the Church were based on ancient writings and beliefs.* Unlike Galileo, most climate skeptics draw conclusions that are not supported by the evidence. Their reaction to climate change is often motivated by ideology, and the arguments that they make from this foundation suffer from one or more errors such as cherry-picking data, confusing weather for climate, and focusing too much on uncertainty. The remaining chapters of this book take on these arguments—not with ideology, but with science.

There is another interesting contrast between climate-change skeptics and Galileo. While today it may seem like the idea that human activity causes changes to the climate is the norm (and therefore that climate skeptics fit the role of Galileo by questioning it), the concept of anthropogenic climate change actually upended the long-held belief that the Earth's atmosphere is a stable and self-regulated system that humans could not affect. Pioneering scientists like Svante Arrhenius and Guy Stewart Callendar, who were among the first to introduce the connection between carbon dioxide emissions and global warming, are the ones who could justifiably compare themselves to Galileo. Climate skeptics are actually pushing to restore an outdated view—and one in stark disagreement with the facts.

To paraphrase Senator Lloyd Bentsen: Brad, you're no Galileo.

---

*There were prominent Catholics, such as some Jesuit astronomers, who repeated Galileo's observations and confirmed his findings, but their points of view were contrary to the official position of the Church at the time.

# There is no consensus

**AS WE DISCUSSED** in the Prologue, science is based on consensus. Consensus does not mean you have a vote with the majority winning. When a scientist puts forward a question—say, "What happens when you throw gigatons of carbon dioxide into the air?"—the scientific community comes up with all sorts of possible answers. Over time, the community tests, and tests, and tests each of these ideas, most of which turn out to be wrong. In science, you know you have consensus when most scientists simply stop arguing against the emerging consensus point of view because the evidence supporting it is too compelling to disagree with anymore.

If you wanted to deliver a knockout blow to something like the idea of anthropogenic climate change, a logical place to start would be to question the existence of a scientific consensus on the topic. If there's no consensus in the first place, there's no need to come up with some complex scientific argument to fight against the consensus. It shouldn't surprise you, then, to learn that Brad has something up his sleeve along these lines.

Brad often receives inadvertent assistance from the general media—especially in the United States—where reporters' efforts to present a "balanced" story on climate change tend to result in extreme minority points of view getting equal time with the consensus view. A consequence of this is that the general public has an unbalanced perception of the facts, and this, in turn, can delay action on mitigating climate change. In defense of the media outlets that propagate this unbalanced "balanced view," there has been a long-standing and well-funded campaign by fossil-fuel lobbies and other special-interest groups to try to convince the public that global warming is just a shaky theory and not the scientific consensus.*

If there isn't consensus, that must mean that there's a boatload of scientists who don't subscribe to the idea that humans are the dominant reason for recent changes in the climate. How does 32,000 sound? Pretty big number, right? That's the approximate number of signatories on a 2008 petition circulated by the Oregon Institute of Science and Medicine (OISM) (the *what?*) that states that there is no convincing evidence that humankind's greenhouse-gas emissions are causing global warming. Brad is feeling pretty smug with 32,000 trained scientists at his back. That is, until you tell him that there are more than *ten million* people in the United States who fit the qualifications laid out by the OISM regarding who can be defined as a scientist (the petition was only circulated in the United States). A little math will reveal that 32,000, as big as it sounds, is well below one-half of 1 percent of the scientist population in one country alone. Not exactly enough to threaten consensus.

More important, though, is that if you dig into the fields of the people who signed this petition, it turns out that there were not many climatologists—only 39, to be exact. If you're generous in your definitions and include ocean scientists and people in other related fields, the number climbs to about 150 people.† This represents a tiny fraction of the active researchers in climate science and related fields—certainly far too few to use as evidence that there is no consensus. Moreover, the organizers of this petition have not revealed how many scientists were polled, nor the sampling methodology that they employed; rather, they only made public the sum-

---

*For lots of examples, see Riley E. Dunlap and Aaron M. McCright, "Organized Climate Change Denial," in *The Oxford Handbook of Climate Change and Society*, edited by John S. Dryzek, Richard B. Norgaard, and David Schlosberg (Oxford University Press, 2011), 144–160.

†Tens of thousands of the signatories were from very non-geophysical science fields, such as computer science, mathematics, and medicine.

mary data of those who agreed to sign. So, among other things, the Oregon Petition didn't seem to ask the right people.

The Oregon Petition was an attempt to collect a list of folks who don't accept that humans drive climate change—not an attempt to determine the actual perspective of the climate-science community. Surely there are organizations that have attempted to take a more complete account of the numbers. Is there consensus after all? Unequivocally yes, and we'll get to that, but first let's allow Brad to fire a few more shots across the bow.

What does the peer-reviewed climate-science literature have to say?* Brad is eager to tell you that the Hudson Institute (a US-based conservative think tank) analyzed journal publications and claims to have dug up five hundred papers refuting the idea that recent global warming is primarily driven by human activity. If you take a look at these papers, however, you'll find this characterization misleading at best, and really just downright wrong. The vast majority of these papers address the Earth's climate history, which, as we discussed in the Prologue, has changed in the past for natural reasons. The fact that global climate change over thousands of years is natural has little or nothing to do with anthropogenic global warming. Other papers discuss ways in which the climate is affected by natural phenomena today, such as cycles of the sun (see Chapter 11). These reports make no claims that humans are not affecting the climate as well. What it comes down to is that the scientific consensus is so broad that most scientists don't bother even to mention it in their articles—it's just assumed that their peers know that humans are the principal driver of recent global warming. As for the extremely small number of papers that do explicitly reject the consensus? We'll cover some of those theories in Chapter 13.

Brad's not done quite yet, though, because it's pretty much

---

*The key word there is "peer-reviewed." This is how scientists vet each other's work to make sure it is accurate. Scientists have to convince their expert peers that their conclusions are sound before the work can be published.

impossible to have a detailed discussion about climate change without at some point having the Intergovernmental Panel on Climate Change (IPCC) enter the conversation. The IPCC was created by the WMO and the United Nations Environment Programme (UNEP) back in 1988 and is the leading international organization assessing climate change, its consequences, and our options concerning what to do about it. Given its prominent role and visibility, the IPCC is the subject of seemingly endless attacks from skeptics like Brad. The subject here is consensus, and, starting in 1990, the IPCC has released assessment reports every few years that provide a comprehensive picture of the latest climate science—an overview of the current consensus view on the subject. Skeptics have claimed that the IPCC does not, in fact, represent the viewpoint of thousands of climate scientists, but is rather the soapbox for just a handful. (This myth touches on conspiracy theory, which we'll get to in more detail in Chapter 2.)

Keep in mind here that the IPCC does not actually conduct research itself—it simply collects and summarizes the research findings of thousands of scientists. The vast majority of these findings come from peer-reviewed scientific journals, and in the cases where information is pulled from other sources, the IPCC has a formal and open procedure for using it. The team that pulls together IPCC assessment reports is large and diverse. For example, for the most recent complete assessment report (2007), there were more than eight hundred contributing authors. (Makes you wonder if the IPCC's Nobel Peace Prize was for peacefully figuring out in what order to put all their names!) Clearly, with so many expert voices involved, the suggestion that the IPCC misrepresents the science is, well, a misrepresentation of the facts.

So what are the facts? We'll go over several studies that have looked into this question, but we'll start with the big gun, in case Brad is on his way to your place right now and you don't have time for the whole story. Probably the most comprehensive attempt to date to review the scientific literature in order to gauge the degree of consensus on climate change is the Consensus Project, organized by a group called Skeptical Science run by a scientist at the Global Change Institute at the University of Queensland. This study examined more than 12,000 peer-reviewed climate-science abstracts published over the twenty-year period from 1991 to 2011. Note that these were not cherry-picked abstracts. There were simply about 12,000 peer-reviewed articles on the subject of "global warming" or "global climate change" during that time—the study looked at all of them. Every one of these abstracts was reviewed independently by at least two international (self-identified) climate experts to determine whether it *explicitly* or *implicitly* rejected/supported the consensus or if it expressed no opinion.* The Consensus Project also received surveys back from the authors of about 1,400 of these papers as to their own view of the position of their paper(s). (Who would know better than the people who researched and wrote them?)

Now, most of the abstracts don't actually take an apparent position because, as we've mentioned before, climate researchers know darn well that humans are causing global warming and see no reason to use some of their precious journal-abstract word limit on something so obvious. Think of it like a chemist taking the time to explain that molecules are made up of atoms in an abstract on the development of a new cancer drug. Or maybe an aerospace engineer spending time on Newton's laws of motion in an abstract about fluid dynamics for a new wing design.

In any case, of those abstracts that did express either an explicit or implicit position, 97.1 percent supported the idea of anthropogenic climate change. Those author surveys? 97.2 percent share the consensus view. That's why it's a consensus! There are also several other independent studies of this kind reported in recent years that

---

*A study of similar scale led by James Lawrence Powell searched only for explicit rejections of the consensus and found a mere 24 out of 14,000 for abstracts published between 1991 and 2012!

all converge on a percentage in the high 90s for how many climate scientists endorse the consensus view. Try getting 97 percent of a group of people to agree about much of anything and you'll see why skeptics who claim there is no consensus on human-caused global warming are just flat-out wrong.

Another interesting result comes out of this project. Since the papers that were analyzed spanned twenty years, the Skeptical Science team could track how the consensus has evolved over two decades. As you've probably figured out, thanks to our less-than-subtle foreshadowing, the consensus has clearly strengthened over time. It was already slightly over 90 percent at the beginning of this period, but as of 2011 it had climbed to 98 percent.

Seems like no further evidence is necessary to make this point, but if Brad can be relentless, we might as well be, too. One way in which scientists get together is through professional organizations and societies. Among the more prestigious of these groups is the UK-based Royal Society. *Climate Change: A Summary of the Science*, a report released in 2010 by the Royal Society, has been highlighted by skeptics for supposedly stating that global warming has stopped (see Chapter 7). This is classic climate skeptic cherry-picking and confusing of timescales.

The skeptic arguments related to this report are largely built around the topic of uncertainty, which is an important topic that we'll get to in a bit, but as for global warming's status, the report does address some specific temperature trends on rather short timescales that in no way refute the consensus view. In fact, here's a telling quote from that very report: "There is *strong evidence* that the warming of the Earth over the last half-century has been caused largely by *human activity*, such as the burning of fossil fuels and changes in land use, including agriculture and deforestation" [emphasis added]. That is essentially a concise affirmation of the consensus view.

So, if the Royal Society doesn't, do any scientific organizations have official positions that do refute the consensus? Not that we know of. Not a one.

Do any endorse the consensus? The answer to that is an emphatic yes! Here's a (probably incomplete) list:

1. Academia Brasileira de Ciências (Brazil)
2. Academia Mexicana de Ciencias (Mexico)
3. Académie des Sciences (France)
4. Academy of Science of South Africa
5. Accademia dei Lincei (Italy)
6. African Academy of Sciences
7. American Association for the Advancement of Science
8. American Astronomical Society
9. American Chemical Society
10. American Geophysical Union
11. American Institute of Physics
12. American Meteorological Society
13. American Physical Society
14. Australian Academy of Science
15. Australian Bureau of Meteorology
16. Australian Meteorological and Oceanographic Society
17. British Antarctic Survey
18. Cameroon Academy of Sciences
19. Canadian Foundation for Climate and Atmospheric Sciences
20. Canadian Meteorological and Oceanographic Society
21. Chinese Academy of Sciences
22. Commonwealth Scientific and Industrial Research Organisation (Australia)
23. Deutsche Akademie der Naturforscher Leopoldina (Germany)
24. Environmental Protection Agency
25. European Federation of Geologists
26. European Geosciences Union
27. European Physical Society
28. Federation of American Scientists
29. Federation of Australian Scientific and Technological Societies

30. Geological Society of America
31. Geological Society of Australia
32. Geological Society of London
33. Ghana Academy of Arts and Sciences
34. Indian National Science Academy
35. International Union for Quaternary Research
36. International Union of Geodesy and Geophysics
37. Kenya National Academy of Sciences
38. l'Académie Nationale des Sciences et Techniques du Sénégal
39. Madagascar's National Academy of Arts, Letters and Sciences
40. National Academy of Sciences (USA)
41. National Center for Atmospheric Research (USA)
42. National Oceanic and Atmospheric Administration (USA)
43. Nigerian Academy of Sciences
44. Polish Academy of Sciences
45. Royal Meteorological Society (United Kingdom)
46. Royal Society (United Kingdom)
47. Royal Society of Canada
48. Royal Society of New Zealand
49. Russian Academy of Sciences
50. Science Council of Japan
51. Sudan Academy of Sciences
52. Tanzania Academy of Sciences
53. Uganda National Academy of Sciences
54. Zambia Academy of Sciences
55. Zimbabwe Academy of Sciences

Enough already! Obviously, the scientific community is settled and, if anything, the consensus is getting even stronger with time. Frustratingly, though, the general public's acceptance of the scientific consensus is far more variable.

Gallup has repeatedly polled the public regarding their views on climate change. They've asked, for example: Which of the following statements do you think is most accurate—most scientists believe that global warming is occurring, most scientists believe that global warming is *not* occurring, or most scientists are unsure about whether global warming is occurring or not?

The results of these polls don't show the same trends as the scientific consensus. The first data points are from around the end of 1997, and they showed that only 48 percent thought that scientists believe global warming is occurring (39 percent thought scientists don't think it's happening, and 7 percent thought scientists weren't sure). These numbers bounced around for a while, settling back on nearly the same result as recently as 2010 (52 percent/36 percent/10 percent). Even though these numbers are lower than we would expect and want, given the fact of scientific consensus, it is encouraging to note that the trend of public perception on this topic has moved increasingly in one direction over the past four polls (the most recent one was in early 2013: 62 percent/28 percent/6 percent). (However, if there's anything we hope you learn from this book, it's that short-term trends can be misleading when it comes to climate change!)

While there are surely all sorts of complex social reasons driving the discrepancy between public perception and scientific consensus, what it seems to come down to is knowledge about climate science. Peter Doran, an earth and environmental sciences professor at the University of Illinois at Chicago, and his student Maggie Kendall Zimmerman published a paper in 2009 reporting the results of their survey of several thousand earth scientists. The scientists were asked if they thought human activity is a significant factor in climate change, and about 82 percent said yes. Doran dug a little deeper* and separated the earth scientists into non-climatologists and climatologists and the latter group into climatologists who publish a lot (a proxy for knowledge on the subject) and those who don't.

---

*Pun totally intended! Much of Doran's research involves studying stuff under the ice in Antarctica.

Now we've got some data on a spectrum of people with progressively increasing knowledge about climate science. Recall that alignment with the scientific consensus among the general public is around 60 percent, and among earth scientists in general it is around 80 percent. Well, the number climbs to 97.5 percent when you look just at the active climate scientists. (Perhaps not so surprisingly, this is almost exactly the same percentage found by the Skeptical Science study of the climate-science literature.) As NBC public service announcements love to say, "The more you know. . . ."

Speaking of what you know, let's turn to the important subject of uncertainty, because it's also tied up in this whole myth about there being no consensus. There are remarkably few things that can be considered 100 percent certain. (Some say only death and taxes, but knowing Brad, he might not even be sure about the latter.) In the world of science, there are laws of motion and thermodynamics and electromagnetism that we're pretty darn sure are reliably true (funny enough, one of these is intertwined with one of the more esoteric climate myths—we'll get to it in Chapter 13). But most science isn't a 100 percent kind of thing. That's why scientists work by consensus. If progress had to wait until every scientist on the planet agreed with every conclusion, things would go nowhere.

Science is often about probabilities. It's about narrowing uncertainties to improve our understanding. The confidence in the fact that humans are affecting the climate has been growing for a very long time. (Remember Svante Arrhenius from the Prologue? He raised the idea back in 1896!) Today, confidence that the planet is warming is very strong (see Chapter 7), as is confidence that humans play a major role in this warming (see Chapters 10 and 11). These points approach 100 percent certainty.* Even so, there are other topics related to climate change where the current under-

*The IPCC 2007 Assessment Report estimates the certainty as greater than 90 percent. Each IPCC report pushes that number higher as climate scientists learn more, with the recently released 2013 IPCC Working Group I report revealing 95 percent certainty. Odds are that the next report will be higher still.

standing is less certain. One of the biggest areas of uncertainty in climate science today, for example, is the role of atmospheric aerosols on global temperature changes. This is the subject of many current research studies.

Climate skeptics jump on the fact that there is this kind of uncertainty in climate science to say that as a society we should take a wait-and-see approach. Why invest money in alternative energy and energy-efficiency technologies if we're not sure that climate change is happening? But the uncertainty about things like aerosols in no way means that there's substantial uncertainty with the big points: that global warming is real and that it's basically our doing.

You can find an up-to-date analysis of the consensus on these big points in a summary statement released by the IPCC related to the upcoming release of its Fifth Assessment Report. The group that authored this summary considers evidence based on peer-reviewed scientific studies and analyses from direct observations, paleoclimate archives, theoretical studies, simulations using climate models, and expert judgment. The degree of certainty in all of these areas is reviewed and updated and reflects the most current consensus of scientific evidence of anthropogenic climate change. In short, this summary states that warming of the Earth's climate is unequivocal. The atmosphere and oceans have warmed, the amounts of snow and ice have diminished, sea levels have risen, and the concentrations of greenhouse gases have increased. While there may be uncertainties about the individual components of the Earth's systems, the only systematic and logical conclusion that can be made is that the changes that international researchers have been seeing in our Earth's climate since the 1950s can only be explained by human influence.

But even if we were to pretend for a moment (for Brad's sake) that there's uncertainty on the big points (which there's not), the fact is that we make decisions based on incomplete information all the time. And the ramifications of not acting in this case could be catastrophic (see Chapter 3). Are we 100 percent sure it'll be catastrophic? No. Are you 100 percent sure that your house won't burn to the ground? No. You have home insurance. We probably don't expect to die young, but we buy life insurance. And the likelihood of climate change wreaking untold havoc is far higher than the likelihood of either of those things happening to you. Consider mitigation of climate change like a global insurance policy, and it's a pretty sure bet that it's the best way to avoid the worst effects of our warming planet.

# It's a conspiracy

**CONSPIRACY THEORIES ABOUND** when it comes to climate change, with some even focusing on the term "climate change" itself. A common skeptic myth is that "they" conspired to change the name from "global warming" to "climate change" because (1) the Earth stopped warming and/or (2) "climate change" seems more scary.* In reality, both terms have been in use by scientists for many years, and they technically refer to different things. Global warming is the phenomenon of rising average temperatures around the planet, which certainly has not stopped (see Chapter 7). Climate change, on the other hand, refers to transformations of the global climate. Patterns of precipitation, heat waves, hurricanes, and the like are examples of components of climate. In the current context, the term refers to changes in climate in response to the increasing average temperature. Scientists have been writing about climate change—and using the term—at least since the 1950s. It isn't new.

There is a new term, however, coined by John P. Holdren, who

---

*Interestingly enough, Republican strategist Frank Luntz composed a 2003 memo to conservative politicians entitled *The Environment: A Cleaner, Safer, Healthier America* in which he advised: "It's time for us to start talking about 'climate change' instead of 'global warming.' . . . 'Climate change' is less frightening than 'global warming.' As one focus group participant noted, climate change 'sounds like you're going from Pittsburgh to Fort Lauderdale.' While global warming has catastrophic connotations attached to it, climate change suggests a more controllable and less emotional challenge." Both terms are scary, if you ask us, but it seems maybe the folks actually conspiring to change terms are from the same camp as many of those accusing the scientists of doing it.

serves President Obama in the following impressive trifecta of roles: Assistant to the President for Science and Technology, Director of the White House Office of Science and Technology Policy, and Co-Chair of the President's Council of Advisors on Science and Technology,* and we generally prefer this term to both global warming and climate change. Holdren has argued that the term "climate disruption" is more appropriate because other terms imply something gradual, uniform, and benign, while what we're experiencing is none of these. You'll find that we use all of these different terms in this book, in some cases to draw on their different meanings, and in other cases because, well, variety is the spice of life.

All right, so whether it's global warming or climate change or climate disruption, "they" didn't conspire to trick anyone with these terms. Often, the "they" in these skeptic conspiracy theories is the Intergovernmental Panel on Climate Change (IPCC), a unique organization. Climate science is the only field of science that periodically appraises every important published result in its field and assembles them all into a giant study. These reports from the IPCC are enormous (and growing), with the Fourth Assessment Report coming in somewhere over three thousand pages. You couldn't be blamed for thinking that there's an army of IPCC employees paid to work full-time on these reports. In reality, the IPCC relies on hundreds of unpaid scientists from all over the world to do the bulk of the heavy lifting. The actual IPCC staff? Twelve people, give or take. Compare that to, say, the World Health Organization and its approximately 1,800 regular staff and $4 billion annual budget.

For comparison, the IPCC budget is in the ballpark of $10 million per year. You read that right. As a society, we are spending about the same on the foremost international organization assembling information on arguably the greatest global threat of

---

*He somehow managed to pull off a step up from his previous trifecta of: Professor of Environmental Policy at the Kennedy School of Government at Harvard University; Director of the Science, Technology, and Public Policy Program at the Belfer Center for Science and International Affairs; and Director of the Woods Hole Research Center.

our time as we do on two or three thirty-second Doritos ads during the Super Bowl. Even if the IPCC were some conspiratorial cabal, with resources like that, they'd be hard-pressed to pull off much of a conspiracy. Let's see what skeptics have claimed that the IPCC has supposedly done.

The king of all climate-change conspiracy theories has its own moniker: Climategate. This story starts in late 2009, when hackers broke into the servers at the University of East Anglia and nabbed loads of e-mails from the Climate Research Unit (CRU). This organization was created in 1972, and it consists of a staff of around twenty research scientists and students. CRU has developed a number of the data sets used in climate research, including contributions to the land surface global temperature record used to monitor the state of the climate system and incorporated in recent IPCC reports. Of special importance for the Climategate story is the CRU's reconstruction of pre-1850 temperatures based on tree-ring measurements.

The hacker(s) behind the e-mail theft were never identified or caught. Norfolk police in the UK wrapped up an investigation into the incident unsuccessfully after three years due to a statutory limitation placed on the investigation by the Computer Misuse Act of 1990. Detective chief superintendent Julian Gregory, the senior investigating officer, did say, however, that "the data breach was the result of a sophisticated and carefully orchestrated attack" and that there was no evidence that anyone working at or associated with the University of East Anglia was involved in the crime.

While several of the thousands of stolen e-mails sent over the course of thirteen years have been highlighted by climate skeptics for their supposedly conspiratorial content, far and away the e-mail most fervently brandished is this one from CRU scientist Phil Jones:

```
I've just completed Mike's Nature trick of
adding in the real temps to each series for
the last 20 years (ie from 1981 onwards) and
from 1961 for Keith's to hide the decline.
```

The two key phrases here are "Mike's Nature trick" and "hide the decline." To a skeptic's suspicious eye, these both seem like pretty damning expressions. Skeptics argued (and the mainstream media subsequently repeated) that these climate scientists were supposedly admitting to manipulating the temperature data to conceal the fact that global temperatures are decreasing. Skeptics are once again cherry-picking here and willfully ignoring the context and the bigger picture, as we'll explain. One point that is clear to anyone is that e-mails with close coworkers are generally casual and brief in nature. In the case of this e-mail message, the recipient was a colleague with intimate knowledge of the meanings of Phil's jargon.

Since presumably none of us is one of those close colleagues of Phil's, let's take a look at what "Mike's Nature trick" and "hide the decline" really meant in turn—and considering them separately is indeed the right approach, because they actually don't have much to do with each other, despite the common mistake made by skeptics of condensing this quote down to "Mike's Nature trick to hide the decline."*

So who the heck is Mike? And what's his Nature trick? Well, Mike is climate scientist and Penn State professor Michael Mann. He was the lead author on a paper published in the journal *Nature* in 1998 in which he plotted recent temperature data on the same graph with historic reconstructed temperature data in order to provide context for the recent warming trend. He put two sets of data together on a graph. Yeah, not much of a trick, really. Would've been way cooler if Mike's Nature trick involved making a bear disappear or something like that.

Phil was using "trick" in the sense of "trick of the trade." It is a common plotting technique in climatology to merge historic data from things like ice cores with data from modern times. And researchers who use Mike's trick, including Mike himself, typically

---

*One such example of the use of this quote is from a public lecture by onetime climate skeptic and Berkeley professor Richard Muller.

# MIKE'S NATURE TRICK

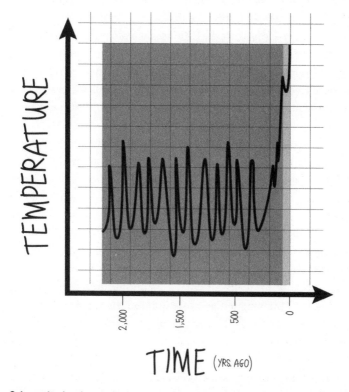

Schematic showing surface air temperature levels, combining historical recon-struction data and more recent direct measurements. The data schematically represented here are often referred to as the "hockey stick" plot, and combining data in this way is "Mike's Nature trick."

do a very good job of clearly labeling their graphs to indicate which part of the data is modern direct measurement and which is historical reconstruction. Think of it like putting together a family photo album with old black-and-white photos and more modern color ones taken with your digital camera to show how the people have aged. This isn't suspicious behavior. No conspiracy there.

So what's this "hide the decline" business? Surely that's a sign of a cover-up, right? Skeptics say that these CRU scientists were conspiring to hide the alleged fact that global temperatures aren't warming after all. It turns out that the decline in question here isn't even referring to global temperatures. The decline that Phil was talking about involves the tree rings that the CRU uses as a proxy for historical temperatures. Growth of tree rings (width and density) correlates well with the temperature of the environment where the tree is growing. Because trees can live for a very long time, this allows researchers, like those at the CRU, to determine what the temperatures were at times before modern measurements began.

Interestingly, looking at the rings revealed that something strange started to happen around 1960 or so. A subset of the trees, those at certain high-latitude (northern) locations, broke from their brethren—they were a rebellious lot—and their tree-ring growth stopped tracking the local temperature. Scientists even have a name for this phenomenon; it's called the "divergence problem," and it has been studied and openly discussed in the literature for years (including in IPCC reports). It really is a surprising situation, because basically all trees—including these troublesome fellas—were tracking temperature quite nicely dating all the way back to medieval times. This divergence problem is the decline that Phil was talking about in his e-mail: the decline in tree-ring growth in these particular trees.

OK, so this raises two questions: (1) What's the deal with these trees that made them stop tracking temperature? and (2) What did Phil mean when he wrote that he wanted to hide this decline in tree-ring growth? (That is, was he tampering with temperature data?)

Well, for the first question, all we have at this point are some educated guesses from researchers, since they're still working on figuring it out themselves. The two most likely causes seem to be air pollution effects* and/or drought effects. Pointedly, these phenomena connect back to global warming. The pollutant sulfur dioxide is a major source of atmospheric aerosols, which impact climate; CFC pollutants are very powerful greenhouse gases; and the increased frequency and severity of droughts is an expected byproduct of a warming planet.

It is common practice to avoid using tree-ring data from the renegade northern trees in calculating temperature trends (since they are now useless for that!), but was Phil (or were his colleagues at the CRU) tampering with other temperature data?

As a result of the media coverage of Climategate and the public uproar that it inspired, a review team, funded by the University of East Anglia but wholly independent from it, was formed to look into this question. The review team was composed of experts on scientific research methodology who had no connection with either the CRU or the IPCC (examples were the vice chancellor from the University of Glasgow, the general secretary of the Royal Society of Edinburgh, and BP's group head of research and technology). These experts found that they were able to replicate the CRU's results using publicly available data and that the CRU's handling of the data did not incorporate any bias (e.g., the CRU did not select only temperature stations that showed more warming).

This is sounding less and less like a conspiracy to cover up data. It turns out that the University of East Anglia's Independent Climate Change Email Review was not the only investigation initiated in response to the Climategate tempest. Pennsylvania State University completed an investigation into Mike of Mike's Nature

---

*Sulfur dioxide is one possible culprit, and lower stratospheric ozone concentration caused by CFC emissions is another. Ozone depletion would result in extra (damaging) UV radiation hitting those trees . . . and they have a heck of a time applying sunscreen.

trick, since he is on the faculty there. This inquiry found "no credible evidence that Dr. Mann had or has ever engaged in, or participated in, directly or indirectly, any actions with an intent to suppress or to falsify data." Separate investigations by the UK House of Commons Science and Technology Committee, US Environmental Protection Agency, Department of Commerce, and National Science Foundation all concurred that there was no scientific misconduct by Mike, Phil, or anyone else involved in the Climategate controversy. Basically, in the great words of Gertrude Stein, there is "no there there."

These CRU guys are not completely off the hook, though. Phil and his colleagues did do something wrong—something of a nature that really stokes the fire of conspiracy theorists. After the hacked e-mails were publicized and the subsequent media storm raged, various skeptics started inundating the CRU with Freedom of Information Act (FOIA) requests. These requests demanded access to raw temperature station data and related confidentiality agreements. It seems apparent that the requests were the product of a campaign intended to overwhelm the CRU. This strategy was revealed in one request, apparently from a participant lacking much attention to detail, for information "involving the following countries: [insert 5 or so countries that are different from ones already requested]."

However, the law does not provide for much flexibility, regardless of the frivolity of FOIA requests. Through a combination of understaffing, lack of familiarity with FOIA laws, and general suspicion of the intent of the requests, CRU staff failed to fully comply with the requirements. This lack of response was then fodder for skeptics to claim—not entirely falsely—that the CRU was intentionally hiding data from public view. That said, as we've explained, there is no evidence that any of the CRU's analysis or contributions to IPCC reports are flawed. And regardless, theirs are but a tiny fraction of the overall body of evidence supporting the existence of anthropogenic climate disruption.

Climategate has simmered down—at least in the general media—as the facts have become clearer now, but skeptics are still stirring the pot in their hunt for conspiracies. Running a distant second place in the contest for biggest supposed climate-change conspiracy is Glaciergate. Unlike Climategate, this one involved an actual mistake made by the IPCC. Skeptics routinely accuse the IPCC of conspiring to exaggerate both the degree of climate disruption and the danger of its effects. In this particular instance, they did in fact exaggerate a particular danger, though there was certainly not a conspiracy to do so. But before Brad and friends get too smug, let's talk about what really went down.

An Indian glacier scientist named Syed Hasnain, from Jawaharlal Nehru University in Dehli, was interviewed for a 1999 article that appeared in the popular science magazine *New Scientist*. In this interview, Dr. Hasnain made a speculative assertion that all Himalayan glaciers might disappear by 2035 if current warming trends continued.* That was mistake #1. This assertion is not supported by any peer-reviewed research. (To learn more about glacier melting due to global warming, check out Chapter 8.) If things had ended there, though, there probably wouldn't be a Glaciergate.

In 2005, the World Wildlife Fund (WWF) published a report on glacier retreat in the Himalayan region, and they repeated the unsubstantiated prediction, citing the *New Scientist* article as their source. Mistake #2. (Note that the WWF report was also not peer-reviewed.) If things had ended there, we still probably could have avoided Glaciergate. But like a juicy piece of office gossip, this assertion just wouldn't go away and wormed its way into yet another report . . . and that's when things hit the fan. This report was the IPCC Fourth Assessment Report. Mistake #3, and a doozy at that. The IPCC cited the WWF report as their source. IPCC reports have strict rules about drawing information from peer-reviewed

---

*He actually made a similar claim in an interview a few months earlier with the Indian environmental magazine *Down to Earth*.

scientific literature (or very clearly indicating otherwise); moreover, these reports are rigorously reviewed at several stages prior to release to hunt down situations exactly like this one. Clearly, the process failed in this instance.

In the IPCC's defense, we're talking about a single statement buried in a three-thousand-page report, but nonetheless, it shouldn't have found its way in there. Lost in all the ensuing firestorm of skeptic and media attention to the IPCC's apparent exaggeration of the effects of global warming* was the fact that the Himalayan glaciers are indeed melting, threatening a resource of vital fresh water for hundreds of millions of people. Alarm is warranted: The glaciers won't completely disappear by 2035, but they are melting at an accelerating rate.

So the IPCC was guilty of overstating one effect of climate disruption in one of their reports, but skeptics argue that Glaciergate wasn't an isolated incident and that the organization routinely conspires to overstate the threat of climate change. What do the facts show? History has shown that the IPCC is remarkably accurate, but when it errs it seems to err on the side of being too conservative in projecting the future (see Chapter 9 for more details). Looking back at previous reports and comparing them with real-world observations since their release, we find that the IPCC has underestimated important things such as the rate of sea level rise, Arctic sea ice melting, and the increase of carbon dioxide emissions. They've even underestimated the degree to which the scientific community has reached consensus that human activity is the most significant driver of recent climate disruption (Chapter 1). This is not an organization contriving to embellish either our role in climate change or the scope and scale of that change's impacts.

We've explained a situation where the IPCC messed up by including one unsubstantiated claim about Himalayan glaciers in

---

*There was even substantial pressure on Rajendra Pachauri, the chairperson of the IPCC, to step down.

an assessment report, but skeptics have attacked the IPCC for several other supposedly conspiratorial incidents as well. And in these cases, it's the skeptics who've made the mistakes.

The first one of these involves the Amazon rain forests. In its Fourth Assessment Report, the IPCC stated that "up to 40% of the Amazonian forests could react drastically to even a slight reduction in precipitation." The reference they cite for this statement is, gulp, a WWF report. ("Oh no!" you say. Not again!) Skeptics jumped on this claim precisely because of the source. Ah, but in this case, the original source cited in the WWF report is not an unsubstantiated assertion from a popular science publication—it's a 1999 peer-reviewed report in the journal *Nature* written by Daniel Nepstad and coworkers. In fact, there are also other, more recent papers in the scientific literature supporting this claim. Unfortunately for the rain forests, this one is correct.

Another skeptic attack on the IPCC takes things to a new level. The subject at hand is the medieval warm period, a stretch of time from about 1100 to 1400 AD* during which global temperatures were, well, warm. In a very early report from the IPCC, there was a figure—a schematic illustration (not actual data)—showing historical temperature changes over roughly the last thousand years. In that figure, it appears that the temperatures during the medieval warm period were higher than those of today. (However, the temperature axis in the illustration doesn't even have numbers on it, since it's only a schematic diagram.) In the IPCC's next report, presto! This figure was gone! Not only do skeptics accuse the IPCC of surreptitiously hiding data, but they also seem to grant the IPCC magical powers—they say the IPCC "disappeared" the medieval warm period because it was supposedly showing that today's temperatures aren't so scary. In the place of the schematic illustration

---

*Appropriately enough, as you'll see, it was during this time that the modern depiction of the wizard Merlin first appears in Geoffrey of Monmouth's *Historia Regum Britanniae*.

was temperature data, and current global temperatures were shown as being higher than during the medieval warm period. Magic! Conspiracy! Nope. The new figure is scientific data—paleoclimatological data to be precise. It was indeed warm back then . . . just not as warm as it is today.

The original schematic figure came from a 1982 book by H. H. Lamb entitled *Climate, History and the Modern World* in which Lamb plotted fifty-year temperature averages ending in 1950 (that is, ending before the current period of warming attributed primarily to human greenhouse-gas emissions). His temperature plots were also from central England only—not global, and they were not calibrated against instrumental temperature data. Anyway, the point is that the IPCC didn't "disappear" the medieval warm period. They just replaced an outdated schematic figure with the latest scientific data.

Enough drama for the IPCC (for now). Let's turn our attention to another favorite target of skeptics: Al Gore's documentary film *An Inconvenient Truth*—known as probably the world's most famous PowerPoint presentation. Regardless of your views of Mr. Gore, one thing is apparent: Overall, his movie is both scientifically accurate and effective at garnering broader attention to the issue of climate disruption. With careful inspection, one can identify a few errors in *An Inconvenient Truth* that we'll explain shortly, but many of the items claimed to be false or alarmist by skeptics are indeed accurate.

So where did Gore mess up? Likely the biggest error was his ascription of the melting of Mount Kilimanjaro's glacier to global warming. Studies have shown that this particular glacier is probably shrinking mostly because of deforestation. In Gore's defense, however, those findings were published *after the movie was made*! Moreover, as we mentioned in the context of Glaciergate, glaciers worldwide are in fact receding on average because of global warming (see Chapter 8). It's just that the Mount Kilimanjaro glacier happens to be one of those shrinking for a different reason.

Gore can't be blamed too heavily for not knowing the results of a scientific study that hadn't even taken place yet, but there is one other point where you can safely say Gore did make a mistake in his film. This is a mistake of attribution. He gives credit to Lonnie Thompson, a paleoclimatologist* and an adviser on the film, for the data showing historical global temperatures, including the recent rise associated with carbon dioxide emissions. These data were actually an amalgamation of Michael Mann's famous hockey stick graph (yup, same Mike) and the CRU's dataset (yup, the Climategate guys). There is nothing wrong with the data Gore showed, but he didn't properly acknowledge his sources.

Gore has also been attacked for claiming that Himalayan glaciers are shrinking. Skeptics point to the Karakoram glaciers, which sit at the border of Pakistan, India, and China, and are growing. But Gore never claimed that *all* Himalayan glaciers were melting (nor that they'd all melt by 2035!), and the science strongly backs up the claim that, overall, Himalayan glaciers are melting just like other glaciers across the globe.

A related attack has to do with ice in Greenland: Skeptics say it's growing and Gore says it's shrinking. The fact is that ice around the border of Greenland is shrinking, like the movie says, but increased snowfall (due to global warming!) is producing ice gain in the interior of the country. Same goes for attacks on Antarctic melting. If you cherry-pick the data, you can find regions where ice is growing, but overall Antarctica is losing ice. (We spend more time on these ice-related issues in Chapter 8.)

One final topic on which skeptics like to criticize *An Inconvenient Truth* is the connection that the film makes between global warming and big storms like hurricanes. Skeptics say hurricanes have always happened and there's no link to climate disruption; they believe these sorts of assertions are a sign of a conspiracy to scare people. In reality, climate science is clear that global warming leads to increased risk factors for certain extreme weather events. We've

---

*A paleoclimatologist studies changes in climate over the entirety of Earth's history.

got a whole chapter (5) on this issue, too. Overall, the film was a darn good summary presentation of the understanding of climate change at the time it was produced.

Skeptics are relentless in their attempts to expose conspiracies related to climate change, but the facts are clear: Climate science is probably the most transparent field in all of science. As we said, there is no other field that rigorously assesses the entirety of the published work in its area on a regular basis and makes the summary of that work publicly available. You could argue that some of this transparency is a product of the microscope put on the field by skeptics themselves, and it seems clear that organizations like the IPCC are even more rigorous in their work than ever before because of all the attention.

Thanks to the equally relentless work of all those climate scientists, we know with high certainty that we are the cause behind rising temperatures and that the effects are going to be serious. It's time to put the conspiracy theories to rest and get on with solving this challenge.

**2**

# We Don't Need to Worry

Who says climate change is such a bad thing?

We're heading toward an Ice Age, so why worry about warming?

There's no link between warming and extreme weather

# 3

# Who says climate change is such a bad thing?

**WHILE THE CLAIMS** of skeptics like Brad can seem outlandish, paranoid, and maybe even deceitful at times, perhaps no skeptic misconception is as dangerous as this one. Brace yourself . . . Brad is going to start with a premise that you wouldn't expect here. He grants you that the planet is warming up (maybe he read Chapter 7 already?) and (gasp) that humans are likely the dominant reason for it. After you've picked yourself up off the floor, you realize that he's got a zinger for you. There is a school of thought in the skeptic community that climate change is, overall, *actually a good thing*. All those tree-hugger stories about looming global calamities? Just scare tactics. Now, we'll grant you that there are folks out there who take things a bit far, sometimes for dramatic effect. Watching Dennis Quaid and Jake Gyllenhaal in *The Day After Tomorrow* run for their lives from a climatic change that's moving faster than Usain Bolt may make for a good movie (or not), but that scenario takes more than a little artistic license. Believe it or not, there are indeed some positive effects that are expected thanks to global warming. We'll go over these in this chapter, but we'll also cover lots of the negative consequences—and these dwarf the handful of helpful ones. As you'll see, there is no question that, in the grand scheme of things, climate change is unequivocally bad for humankind. Really bad.

Before we delve into the sorts of things that will happen (or in many cases are already happening) as a result of anthropogenic climate disruption, there are a few general concepts to go over. A key thing to keep in mind when it comes to effects of climate change is that they will all be proportionate to the amount of warming. Each additional degree of warming beyond the baseline global temperature will not only introduce new effects, but it will also make each effect more severe. It may seem that one single, lowly degree shouldn't even make a difference. Who among us can really tell the difference between, say, seventy-four and seventy-five degrees Fahrenheit? (No doubt some of you are now turning to argue with a significant other or roommate about that thermostat setting in your home.) Surprisingly enough, that one degree really does have a colossal impact on the planet and its inhabitants.* This is why it's so important to limit our impact on the climate as much as possible. We are already at a point where we've begun to screw with the system, but that doesn't mean we're all doomed. Every bit that we can lessen our greenhouse-gas emissions will make the effects less severe.

Before we send you cowering under your covers and buying emergency kits, let's be fair to Brad and spend a little time talking about the handful of positive effects anticipated to result from a warming planet. The positive outcome that probably receives the most attention is the fact that the Arctic Ocean's Northwest Passage will eventually be free of sea ice (at least in the summer months), thereby opening up a shortcut for marine shipping between the Atlantic and Pacific Oceans. Assuming that the passage was safely navigable (a big "if," since iceberg threats will remain†), it would represent a substantial boon for the shipping industry. A related opportunity would be the ability to lay fiber-optic cables under the Arctic Ocean, which could significantly increase global communi-

---

*The "degree" we're referring to in the context of climate change is really a degree Celsius, which is about twice as large as a degree Fahrenheit. But could you tell the difference between seventy-four and seventy-six degrees Fahrenheit, in any case?

†Perhaps Leonardo DiCaprio and Kate Winslet will return for a remake of *Titanic*, except somehow shipping containers don't have the same romantic mystique as grand ballrooms.

cation speed. Also, there are probably undiscovered natural gas and oil that will become accessible, though of course that's the stuff that got us into this situation in the first place.

Another consequence of maintaining a relatively cooler climate is pretty straightforward: winter kills. Segments of the population that are especially vulnerable, such as the elderly, can succumb to extreme cold weather; every year there are many thousands of deaths worldwide that are attributed to this phenomenon. If the planet warms, you'd expect to have fewer such deaths.

Beyond these effects, there is only one other major outcome purported to be a good thing to come from climate change. The argument is that global warming is good for agriculture and plants in general. For one thing, some regions will likely have longer growing seasons due to shorter winters (though in many cases, these regions receive little sunlight during the extended season anyway because they're at high latitude, so the benefits would be minimal). Another aspect related to plants is that plants absorb carbon dioxide. In effect, it is plant food, so more carbon dioxide in the air means plants have more to eat. Sounds like a good thing, right? Well, it isn't as simple as that.

Brad points out that people have shown that pumping extra $CO_2$ into a greenhouse leads to faster growth of the plants inside. True enough. But hidden in these studies is the fact that these plants have luxurious environments where they receive all the water and nutrients they need to sustain that faster growth, as well as protection from insects. Outside the protective walls of a greenhouse, plants do not necessarily have access to these resources, which means that they may not benefit from having additional food. Water, in particular, is an issue. As the world warms up, evaporation increases, pushing the water needs of plants up even higher. Many agricultural regions are already under water stress, and climate change will have a dramatic impact on not only the demand side but also the supply side of the equation. Even where you might think you'd catch a break, such as in the fact that rainfall is expected to increase in some areas because of stronger storms (see Chapter 5), in reality these heavy downpours cannot be fully absorbed by farmlands, and the runoff carries away both soil and fertilizer.* Agriculture will be hit hard by other factors, too, but let's get back to the plant-food myth for now.

As scientists have researched the topic of carbon dioxide's effect on plants with greater rigor, the focus has shifted from greenhouse studies to so-called free-air $CO_2$ enrichment (FACE) studies. In FACE studies, the experiments are conducted outdoors with a carefully controlled release of carbon dioxide, thereby providing a more realistic replication of how plants would react in a real-world setting with a particular concentration of $CO_2$ in the air. The results from this more sophisticated research are far less promising than those from the greenhouse work. The reasons for the disappointing yields, despite the presence of increased food, range from carbon dioxide making the plants more vulnerable to insects, to changes in soil chemistry that render plants more susceptible to disease, to differences in the way roots develop in the two environments (greenhouse vs. outdoors). Another issue is that some staple crops, such as wheat, have less nutritional value when grown with excess $CO_2$, suggesting that even if yields increase, the added nutritional value would be minimal.

OK, so there are in fact a few good things that might happen as a result of climate change, although not as many as claimed by Brad and friends. Now let's turn our attention to the flip side. Many of the negative impacts revolve around what will happen to animals and plants as the planet heats up. Though it probably irks some of Brad's like-minded brethren, some of whom, in all likelihood, are also evolution skeptics, he tells you that there's no reason to fret about all those creatures and plants because they can simply

---

*Speaking of fertilizers, guess where those come from . . . fossil fuels! Natural gas is the primary source of hydrogen used to manufacture commercial fertilizer, so its use competes with, and encourages increased production of, one of our major sources of energy.

adapt to the changing environment. It turns out that we humans are so adept at messing up ecosystems through centuries' worth of carelessness (agriculture, mining, logging, overfishing, relocation of species, and so on) that it's challenging to dissect which impacts can be traced specifically to anthropogenic climate change. Scientists who would be best qualified to figure that kind of thing out, like conservation biologists, are rightfully more focused on all those other shorter-term threats to ecosystems. That said, there have been some important studies looking at the effects of global warming on various species. The upshot is that the pace of climate change spurred by human activity is so fast (and accelerating!) that species are not well equipped to adapt in time.

We are poised to compress an amount of global warming that historically has occurred over the course of thousands of years into a single century. The most optimistic scenarios project about two degrees Celsius of warming this century, which would put us at a temperature last seen during the Pliocene Epoch, about three million years ago. A more realistic number seems to be around four degrees Celsius. While four degrees doesn't seem like much, you'd have to look back thirty-five million years to find the last time Earth witnessed such a temperature average. That was the Eocene Epoch, and creatures like *Campanile giganteum*—a sea snail about two feet long—and *Gigantophis*—a snake more than thirty feet long—were moseying about the planet. Yikes. Those species, like most that were around back then, are gone now (we'll pause while you sigh in relief). Species generally only stick around for a few million years, though there are plenty of exceptions. The point is that the climate changes that we are in the midst of have never been encountered by most of the species alive on Earth today. Moreover, the stresses caused by climate change are just one part of a collection of stresses caused by human activity, meaning that the buffer for adaptation is basically exhausted already. Will plants and animals be able to adapt, as Brad claims? Perhaps a bit here and there, but in the big picture, the answer is a very unfortunate no. Scientists anticipate extinctions on a frightening scale, with each successive degree of warming adding considerably to the roster of species lost forever. Polar bears are the poster child for this phenomenon, but they are essentially canaries in the coal mine. The effects of climate change are often strongest in polar regions, so the effects there are harbingers of what's to come for the rest of the planet.

The species on Earth of greatest interest to us is, well, us. For that reason, we'll spend the rest of this chapter viewing the effects of climate change through the lens of how they will impact human society. Aspects of our lives that will suffer the consequences of a warming planet run the gamut from water to food to health to the economy. In other words, pretty much all the stuff that is important to our survival as a species and a society will feel the strain. Let's start with water.

We'll begin by introducing you to global warming's sinister cousin: ocean acidification. As we went over in the Prologue (and will cover in greater detail in Chapter 10), carbon dioxide emissions from human activity are a major driver of global warming because they reside in the atmosphere and act as greenhouse gases. Not all of the $CO_2$ stays in the atmosphere, though. Significant amounts of it end up being stored in other places known as "carbon sinks."* Soils and plants soak up huge amounts of carbon dioxide, but the largest carbon sink available on the planet is the oceans. In fact, the amount of $CO_2$ dissolved in the oceans has risen by about 30 percent since pre-industrial times. So what happens when you dissolve carbon dioxide in water? Well, a little chemistry lesson is in order. When $CO_2$ and $H_2O$ mix, you end up with something called carbonic acid: $H_2CO_3$. In other words, carbon emissions that end up in the ocean actually make the ocean more acidic. Remember the old pH scale from science class? Quick refresher: It goes from 0 to 14, with 7 being

---

*One of the ideas on the table to lessen the impact of climate change is to try to capture, or sequester, carbon dioxide artificially and store it indefinitely as compressed gas or locked up in minerals and rocks. Unfortunately, these technologies are not ready for prime time yet.

neutral. The lower the number, the more acidic. (The oceans' pH is just above 8.) That 30 percent increase in carbon dioxide in the oceans has correlated with a drop in pH of about 0.11 so far.

There is a long time lag between the time of emission and the transition of the gases into a carbon sink. Even if we stopped emitting carbon dioxide completely today, the pH of the oceans would still drop about another 0.1 units thanks to the emissions that are already in the air. Who cares about the oceans being more acidic? What are the costs of a more acidic ocean? That carbonic acid that forms in the water from dissolved $CO_2$ scavenges carbonate ions from the ocean, and lots of sea creatures need that carbonate to grow calcium carbonate, which they use in their exoskeletons and shells. Decreasing carbonate-ion concentration means that a broad range of organisms such as phytoplankton, algae, corals, and shellfish have a much harder time growing exoskeletons or shells. These effects are already visible in the erosion of exoskeletons among many creatures at the bottom of the ocean food chain (thereby affecting essentially all marine animals), and the oceans are only going to become more acidic with time. For those who consume fish and seafood, this means a smaller supply and, therefore, higher costs.

A threat to our food supply is plenty of reason for alarm, but the ramifications of ocean acidification don't end there. Coral reefs are among the most biodiverse places on Earth, and they are already succumbing to severe stresses caused by human activity. Corals actually end up suffering both from the carbonate shortage (evil cousin 1) *and* from higher temperatures (evil cousin 2). The death of coral is—somewhat euphemistically—referred to as "coral bleaching," a more palatable term than something more appropriate, like "coral obliteration."

Let's shift from chemistry to a biology lesson. Coral reefs are a symbiotic system composed of coral polyps and algae called zoo-xanthellae (that one might win a national spelling bee one day!). When something interferes with the symbiotic relationship between these two organisms, the polyp kicks the algae to the curb.

The algae are the source of the brilliant colors of coral reefs, which is why this phenomenon is called bleaching. So what sort of thing might make the polyps kick the algae out? It's not cheating with their best friend, if that's what you were thinking. The oceans are warming up (see Chapter 7), and when things warm up, chemical reactions happen faster. One such effect is that warm water causes those zooxanthellae to produce too much waste oxygen, effectively poisoning the coral, but coral polyps don't go down so easily and end up ejecting the algae. Unfortunately for the polyps, those algae provided almost all of their food, so the coral typically starves to death or succumbs to various infections or other unfortunate fates due to its weakened condition. So, in tragic style that would make Shakespeare proud, the coral ends up dying anyway. The combined effects of pollution, global (ocean) warming, and water acidification are causing about 1 percent of coral worldwide to die each year.

Brad scoffs at the thought, saying he's not a recreational diver, so why should he care? For one thing, reefs provide sustenance and livelihood for more than half a billion (that's with a "b") people who fish and who live primarily on seafood. Goods and services derived from coral reefs are estimated to be valued at hundreds of billions (another "b" there) of US dollars per year. Reefs also provide shelter to about a quarter of the fish in the ocean and millions of other species. And the algae on reefs perform photosynthesis, which is the base source of food for the tropical and sub-tropical marine food chains. Basically, they are a crucial foundation of ocean—and ultimately global—health. So yes, Brad, you should care.

Another highly visible effect of climate change is the melting of glaciers (see Chapter 8). You may have seen photographs of glaciers, taken over the course of time, showing dramatic receding of the ice. The reason that this is a cause for serious concern is that about a sixth of the people on the planet depend on fresh water coming from the (natural) springtime melting cycle of glaciers. Once the glaciers shrink enough, there will be insufficient water for consumption or agriculture on a massive scale.

So where does all that water go when it melts away from the glaciers? Into the sea, eventually, and that in turn contributes to another major consequence of climate change: sea-level rise. It turns out that there are several factors involved in raising the sea level, one of which is the melting of ice from glaciers and ice sheets. An even larger contributor to sea-level rise (to date) is a bit surprising. The oceans are literally getting bigger. When things warm up, they expand. (You can prove this to yourself with a filled balloon and a hair dryer—woohoo . . . physics lesson! We hit the science hat trick—chemistry, biology, and physics!) The oceans are warming, so they are also expanding, and that expansion manifests itself as a rise in sea level. (Over time, ice melting will dominate this expansion effect.) Now, if you happen to have coastal property (or are one of the hundreds of millions of people who live along a coast), this is quite clearly a problem for you.* But it's also a problem for the rest of us, and not just because we all in effect subsidize the insurance policies for those coastal properties. Sea rise threatens coastal wetlands—another ecosystem with tremendous biodiversity—which serve as protective barriers against the damage caused by hurricanes. Saltwater from a rising sea contaminates freshwater rivers and aquifers. It also inundates rice paddies in low-lying geographic regions. That threatens food and water that we desperately need.

Beyond the effects on fish/seafood and rice, climate change causes plenty of other negative effects on our food supply. Agriculture requires a stable supply of water. Too much and too little are both bad, and meteorological changes induced by global warming will lead to both situations occurring more frequently and with greater intensity than in the past. Droughts and floods wreak havoc on farming, sometimes destroying entire harvests with a single blow. These events drive food prices into the stratosphere. Wildfires are also becoming more frequent, in large part because

---

*The Maldives government famously held a cabinet meeting underwater to highlight the effect of global warming on their island nation.

of the confluence of higher temperatures, increased drought conditions, and earlier snowmelt. Together with shifts in water supplies, timing of seasonal changes, and encroachment of shrubs on grasslands, these changes are impacting the health of livestock and their food supplies as well. Moreover, while some regions at higher latitudes may become usable as farmland as the planet warms, these regions receive little sunlight and have poor soil for agriculture. The corollary is that regions such as the sub-Sahara and tropics that were already near the temperature at which crop yields begin to fail because of heat stress will cross that threshold as global warming proceeds, resulting in the acceleration of desertification, which will render huge tracts of land* unusable. Together, these effects will mean higher food prices for those of us who can afford them and hunger or starvation for those who can't.

Human health will experience impacts from climate change well beyond hunger. As Brad was eager to point out, winter deaths will likely decline as a result of global warming. Those same groups that were vulnerable to the cold, however, are also vulnerable to heat, and projections are that heatwave deaths will trump the decline in winter deaths by a factor of about four. Seems a fool's bargain to accept that trade. A warmer climate will also spread insect migrations into new regions, and insect-borne diseases such as malaria, dengue fever, and West Nile virus will spread along with them. In fact, this is already happening. But don't fret . . . it's not all death and destruction: Another outcome will be increases in allergic reactions because of a rise in pollen counts.

Economically speaking, climate change is an unfettered disaster. Effects like those we've discussed in this chapter will disrupt financial markets, insurance systems, global trade, energy supplies, and more. Globally, economic losses will surely be measured in many trillions of US dollars (that's with a "t"), with projections in the ballpark of 400 trillion dollars (or 0.4 quadrillion dollars, with a

---

*Not the Monty Python variety.

"q" . . . don't get to use that one too often). Regrettably, developing countries will likely be hit the hardest. These are the countries least prepared to adapt and least responsible for the root causes of climate change in the first place. Many are already entangled in conflict, and ever-decreasing water, food, and energy resources will only exacerbate these conflicts. Such geopolitical instability is a daunting challenge for us all.

One last interesting but alarming consequence of global warming is more global warming. There are many feedback systems built into climate change that lead to an acceleration of the process: Warming begets warming. For example, the melting of sea ice isn't just bad for polar bears and seals. Anyone who has gone skiing on a sunny day can tell you that snow and ice are extremely reflective of sunlight. The sea ice reflects a substantial amount of the sun's radiation back into space. (Scientists describe ice as having a high albedo.) When sea ice goes away, that radiation is absorbed by the planet instead, leading to increased warming, meaning more sea-ice melt, and the cycle begins to run out of control by what is called positive feedback until all the sea ice is gone. (In this context, positive feedback is negative!)

Another feedback system involves the thawing of tundra and melting of frozen sea beds, both of which release enormous amounts of trapped methane (natural gas) into the atmosphere. Methane can be useful as a fossil fuel and for the chemical industry as a feedstock, but in the atmosphere it acts as a powerful greenhouse gas—twenty times more powerful than carbon dioxide. So thawing/melting in polar regions releases greenhouse gases, which further warm the planet, which further thaw/melt the polar regions. You get the picture.

Economists have calculated the cost for the thawing of the permafrost beneath the East Siberian Sea, and it alone has been estimated at sixty trillion dollars (that's with a "t")—a figure comparable with the size of the entire global economy.

# CATEGORIES OF MAJOR IMPACTS DUE TO CLIMATE DISRUPTION

OCEAN ACIDIFICATION

AGRICULTURE

EXTREME WEATHER

SPREAD OF DISEASE

DAMAGE TO ECONOMY

WATER SUPPLY

LOSS OF BIODIVERSITY

SEA LEVEL RISE

**Climate disruption will have far-ranging negative impacts that should have us all concerned (see more on extreme weather in Chapter 5).**

What this all comes down to is that climate change is bad news. Most of these changes are gradual, but many are also frighteningly severe. As in the old boiling-frog story,* we tend not to react to a gradual change even if the consequences of that change are grave. The good news is that there are things we can do to minimize the impacts of climate change. In Chapters 14 and 15, we'll discuss the skeptic attacks on these opportunities and provide the facts about why those attacks are unfounded.

---

*The story goes that if you place a frog in boiling water, it will jump right out. But if you put that same frog in cool water and slowly heat it to boiling, it will never jump out and will obviously then suffer a most unpleasant death. Experiments have shown that this isn't actually the case, but the story makes a good point.

# 4

# We're heading toward an Ice Age, so why worry about warming?

**IN CHAPTER 6**, we'll address skeptic myths that are rooted in the mistake of thinking that short-term, local effects are meaningful in the context of climate change. In this chapter, we first investigate another mistake, this one involving timescales in the opposite direction. Despite multifaceted evidence that we are experiencing a period of prolonged global warming (see Chapter 7), some skeptics argue that we are actually about to enter another Ice Age. If such a climate shift were imminent, the planet would be about to cool dramatically, and so of course we needn't bother ourselves with worrying about carbon emissions.

Why would skeptics even think to make a case for global cooling when all the data point to warming? It starts with those ice cores that we introduced back in the Prologue.

Thanks to deep ice cores from Antarctica, we have a reliable record of global climate dating back hundreds of thousands of years. During that span of time, there is a pretty clear pattern of long Ice Ages, lasting about 100,000 years each, punctuated by relatively brief (11,500 years long on average) warmer periods called interglacials. This pattern is driven largely by astronomical phenomena related to changes in Earth's orbit and tilt. (We will discuss these phenomena, called Milankovitch cycles, in more detail in Chapter 10.) These changes affect the amount of sunlight hitting the Northern Hemisphere. When there is less sunlight, ice doesn't melt as much during the summer. Over time, the ice sheets continue to grow, and because ice is very reflective, more and more sunlight is bounced back into space rather than being absorbed by the planet. This positive feedback cooling cycle continues over thousands of years and eventually spawns an Ice Age.

We currently live in one of those short interglacial periods. In geological terms, this period is called the Holocene Epoch, and it began about 11,700 years ago. That epoch saw the onset of the Neolithic Revolution, a time of wide-scale transition among humans from a lifestyle of hunting and gathering to one of agriculture and settlement. This shift spurred population growth, and the Holocene encompasses the growth of the human species worldwide, including all of its written history through to the present. Ice-core data tell us that the Holocene interglacial period has already lasted several centuries longer than the average interglacial period. This fact has led climate skeptics to believe that the next Ice Age is looming in the near future. Brad says there's no point doing anything to stop global warming with an Ice Age on the horizon.

In reality, there are good reasons to believe that our next Ice Age is a long way off.* Earth's orbital cycles are not as simple as a pendulum swinging back and forth. There is complex interplay among all the various changes in tilt and orbit, and where we currently sit in the cycle is in a time of cooling, albeit rather weak cooling. The last time the planet was in a similar situation was about 430,000 years ago, and the interglacial at that time lasted roughly three times longer than average. Ignoring human changes to climate, and going by what happened last time, we would not expect an Ice Age to kick in for another 18,000 years or so. Even Brad has to admit that this doesn't count as imminent.

However, humans are changing the climate (see Chapters 1, 10, and 11). Warming from our activity—primarily carbon dioxide emissions—is much stronger than the natural cooling from changes in Earth's orbit and tilt. It's also stronger than cooling from recent falloffs in radiant output from the sun. Exactly when the natural

---

*If only the same could be said of the next *Ice Age* movie sequel.

cooling would be strong enough to overcome greenhouse warming from our activity will depend on how much more $CO_2$ we put into the atmosphere. Models suggest that 1,000 billion tons of anthropogenic $CO_2$ would hold off an Ice Age for about 130,000 years. That sounds like a lot of carbon dioxide (well, it is a lot), but we're already about a third of the way there. Sometime, a long, long time from now, a new Ice Age will likely emerge, but we're worried about the next 100 years (see Chapter 3)—not the next 100,000 or more.

Omens of an impending Ice Age aren't new. Mainstream media caught the Ice Age bug back in the 1970s. A widely read *Time* magazine article from June 24, 1974, called "Another Ice Age?" summarized the meme: ". . . when meteorologists take an average of temperatures around the globe they find that the atmosphere has been growing gradually cooler for the past three decades. . . . The weather aberrations they are studying may be the harbinger of another ice age." Similar thoughts were echoed in an April 28, 1975, article in *Newsweek* called "The Cooling World." These articles, and the other media stories they germinated, were the result of journalists catching wind of a few papers in the scientific literature in which a cooling trend was reported. Back then, climate change was not nearly the hot topic it is today, and the number of research studies on the subject reflected that status. In the two decades preceding the *Time* article, there were fewer than ten papers predicting global cooling. On the other hand, there were already dozens predicting warming due to human carbon dioxide emissions. (In the years since that period, consensus among climate scientists that human activity is driving global warming has strengthened dramatically, as we discussed in Chapter 1.) Who knows why the journalists opted to report only on the cooling studies? Maybe Ice Ages sell better? As for why Brad is still clinging to this outdated story, maybe he just misses watching *All in the Family* and playing Space Invaders.

The reason that some meteorologists were projecting a cooler planet in the future back in the 1970s is that things actually had been cooling for several decades. Mid-twentieth-century cooling was driven by human pollution of a different kind: sulfate aerosols. We'll get into this story more in Chapter 7, but what happened next is that governments realized the damage that sulfate aerosols (and other air pollutants) were causing. This eventually led to the passage of pollution prevention laws like the Clean Air Act. When these laws took effect, aerosol levels rapidly began to fall, and their cooling influence left with them, ushering in the current period of global warming.* If only we would enact analogous legislation today to limit $CO_2$ emissions, we could again mitigate humans' impact on the global climate.

So it's clear that we're not heading into an Ice Age anytime soon, but skeptics put another spin on this topic. This myth relates to a period with the adorable name "Little Ice Age." This was the period from the sixteenth century to the nineteenth century, and it wasn't really an Ice Age, though it was certainly a cooler period—at least in the Northern Hemisphere. Farms and villages in the Swiss Alps were severely damaged by growing glaciers in the 1600s, and in the winter of 1780, New York Harbor froze, enabling people to walk from Manhattan to Staten Island! Interestingly, Antonio Stradivari, the famed violin maker, constructed his instruments during the Little Ice Age, and it's said that colder temperatures likely made the wood that he used in his violins denser, contributing to their reputed tone.

Skeptics claim that the planet has been warming up ever since the end of the Little Ice Age, and that this means humans have played no role in recent warming.

The skeptic argument is predicated on an assumption that the Earth naturally bounces back and forth in temperature and that the natural influences that led to the Little Ice Age are now reversing their effects and warming things up. Scientists have a pretty good handle on the natural forcings†—as well as the human ones—that

---

*Warming from the greenhouse effect had been there all along, but it was largely masked by the aerosol cooling effect.

†Forcings are influences (either natural or anthropogenic) on the global temperature.

drive climate changes. The planet doesn't warm up just because it wants to (or because it was cold, like in the Little Ice Age)—there has to be a forcing that makes it warm. Let's examine what likely led to the Little Ice Age, which will help us determine if those forcings can explain recent warming trends.

A major contributing factor to the Little Ice Age was a drop in solar activity. In fact, around that time there were two periods in which solar activity was lower than at any other period in at least a thousand years. These periods are called the Spörer Minimum (1460–1550) and the Maunder Minimum (1645–1715), named, respectively, after German astronomer Gustav Spörer and English astronomer Edward Walter Maunder. (Both studied sunspots, which are a good indicator of solar activity.) A third, smaller drop in activity (the Dalton Minimum, named for English meteorologist, chemist, and physicist John Dalton) took place toward the tail end of the Little Ice Age. That less energy was reaching Earth from the sun certainly goes a long way toward explaining why things were so cold.

If this same forcing (the sun) is to explain our current warming, it would have to mean that the sun's activity is now higher than at any point from the Little Ice Age up to today. Increased solar activity after the Dalton Minimum in the mid-1800s does, indeed, help explain what ended the Little Ice Age. Human industrial activity began a period of rapid expansion prior to the Dalton Minimum, but because the atmosphere is so vast, human influence on our climate has only become a dominant factor since the second half of the twentieth century. So, changes in solar output did play a warming role in the recent past—for some time more powerful than human influences—but what about the most recent period of warming? Has the sun's warming trend continued? Actually, the sun's output has been slowly dropping for the past thirty years or so. Clearly, this particular forcing is not the explanation for current global warming.

Other contributors to the Little Ice Age were a stretch of increased volcanic activity and a decline in human population.

Volcanic eruptions spew sulfur dioxide into the atmosphere, and this leads to the formation of sulfate aerosols; these tiny particles in the air—just like sulfate aerosols from pollution—have a cooling effect because they scatter sunlight that would otherwise reach the Earth. The population drop was caused by the Black Plague, which led to decreased agriculture and, therefore, increased forestation. All those new trees eventually sucked up extra carbon dioxide from the atmosphere, thereby weakening the greenhouse warming effect. What role do these two effects have on recent warming? Volcanic activity, as usual, has had a net cooling effect since 1950, so that can't be why things are warming. As for human activity, well, this is in fact what's driving warming today. Part of this warming is because of our deforestation activities, but a far bigger share of it is because of our $CO_2$ emissions (see Chapter 10).

In other words, while it is true that the planet has warmed considerably since the Little Ice Age, there is no such thing as "recovering" from an Ice Age. Climate changes for a reason, and it's changing the way it is today because of us.

# · 5 ·

# There's no link between warming and extreme weather

**IT SEEMS THAT** every time there's a devastating hurricane or a major flood or drought, some people say it's the result of global warming, while others—like Brad—say we've always had severe hurricanes and floods and droughts and we always will, so climate change has nothing to do with it. The truth is that it's impossible to attach the blame for any single extreme weather event to climate disruption. Weather is inherently chaotic and governed in part by randomness, like a hand of blackjack. But that doesn't mean that there's no connection between extreme weather and climate change. By warming up the planet with our carbon dioxide emissions, we're stacking the deck against ourselves. We're influencing the risk factors for these extreme weather events, and this will result in a trend of more extreme weather over time. In this chapter, we'll explain the mechanisms underlying this connection and discuss some of the evidence that changes are already happening.

It takes a long time to accurately detect trends in rare and highly variable events like major weather occurrences, but climatologists expect global warming to affect extreme weather for at least three reasons:

1. A signature of global warming is warming in the lower atmosphere. Warmer air can hold more humidity—that's why it feels so dry in the dead of winter and so oppressive on a muggy summer afternoon. If there is more water vapor in the air, precipitation—including extreme precipitation—

is more likely. (While precipitation will increase overall on the planet, some regions will see more and others less.)

2. A warmer planet results in faster evapotranspiration. Other than being a difficult word to pronounce, evapotranspiration is, as it sounds, a combination of water evaporation (in this case from the surface of the planet—soil, bodies of water, and the like—to the atmosphere) and plant transpiration. Transpiration is the movement of water within a plant and the subsequent escape of water vapor through its leaves. If the rate of evapotranspiration increases, droughts will be more common and more intense. (Yes, this too will be more severe in some regions than others . . . climate disruption, like politics, is local.)

3. Global warming affects the oceans at least as much as it affects the land and atmosphere. Changes in the temperature of the sea surface can affect patterns of circulation in the atmosphere. These changes can, in turn, lead to droughts and extreme precipitation. Warmer water also provides energy to storms. It's like amphetamines for a hurricane or tropical storm.

We know that the trigger for each of these mechanisms—global warming—is happening (see Chapter 7). The basics of these mechanisms connecting warming to extreme weather are pretty straightforward, though as we said, weather is a complex thing and climate scientists are still working hard on understanding the specific impacts of climate disruption on various types of extreme weather. So the question seems more a matter of when, not if, climate disruption will create more extreme weather.

The short answer to that question is that it's already started. Perhaps the most obvious weather effect of global warming is warming itself. Heat waves have been more common in recent decades, and the chances of record-breaking highs will continue

# THREE MECHANISMS CONNECTING WARMING TO EXTREME WEATHER

MORE HUMID AIR

FAST EVAPOTRANSPIRATION

WARMING SEA SURFACE

to increase as climate disruption proceeds. Our disruption of the climate has already doubled the probability of extreme heat events like the record-breaking summer of 2010 in Russia and 2011 in Oklahoma and Texas.

Prolonged periods of record high temperatures are not just dangerous; they are also a major contributing factor to droughts. Southern Europe and West Africa have experienced more droughts in recent decades. Increased evapotranspiration and regionally decreased precipitation have led to widespread drying, if not droughts, in other parts of Africa, East Asia, and South Asia. These dry conditions in turn can cause more frequent and larger-scale wildfires. While other factors (like poor forest management) also play a role in the occurrence of wildfires, in recent decades there has been a marked rise in wildfire activity that's linked to global warming.

Clearly, these changes vary widely across the globe, with some regions hotter and some cooler. Regional variation is seen with changes in precipitation patterns as well, with some regions getting extra precipitation and some getting less. Recent studies have connected global warming to intensification in heavy precipitation events over about two-thirds of the land in the Northern Hemisphere. The United States Global Change Research Program

(USGCRP), a federal program that coordinates global-change research across thirteen government agencies, compiles some of this information for the United States. According to their latest report (2013), winter storms along the West Coast and New England coast have increased in frequency and intensity. Very heavy precipitation has increased in many parts of the country, with the largest increases in the northeast, midwest, and Great Plains. Heavy downpours have overwhelmed the capacity of our infrastructure, like storm drains, and have produced flooding and accelerated erosion.

Even though certain climate changes are relatively local, they can still have far-reaching influence on the weather, and on much else besides. For example, the decline of summer sea ice in the Arctic (see Chapter 8) increases the amount of heat that the Arctic absorbs. A warmer Arctic affects weather patterns to the south. Dislodged Arctic sea ice floats south into the North Atlantic, where it melts. The resulting relatively fresh water alters the regional ecology and influences the worldwide circulation of the oceans. Changes over the past few decades have had profound consequences, including a transformation of the sea life in the waters around Greenland. Where cod once thrived, now there are shrimp.

For those who don't fish the waters off Greenland, a perhaps more pertinent example of a ripple effect occurs when a massive storm hits the US Gulf Coast. The entire country is affected, because fossil-fuel production and distribution are disrupted. Yet another example is that wildfires generate particulates in the atmosphere that not only can lead to poor air quality a long way from the fire but also can influence the circulation of air, which itself can then produce anomalous weather patterns.

Thanks in part to Al Gore raising the subject in *An Inconvenient Truth* (see Chapter 2), the most debated extreme weather phenomena in the context of climate disruption are hurricanes.* The disproportionate attention on hurricanes is warranted; these are by far the costliest storms in the world. As with all weather events, no single storm can be connected directly to climate change, but is there any connection to be made between these storms and global warming? Let's look at where hurricanes come from.

Two key ingredients in every hurricane are warm water and moist warm air—that's why hurricanes originate in the tropics. Hurricanes are born from warm, humid air near the sea surface. This warm air rises, and the work it does rising up through the atmosphere cools it and causes its vapor to condense into raindrops and storm clouds. This condensation releases heat, which warms the cool air above, causing it to rise and make room for yet more warm, moist air from the surface of the ocean. This cycle transfers heat from the sea surface to the atmosphere, which creates a wind pattern that spirals around a comparatively calm center (the "eye") akin to water swirling down a drain.

We know that sea-surface temperatures and marine-air temperatures have been rising, so the essential ingredients are there for making strong cyclonic storms. There are really two separate questions here: (1) Does global warming cause more hurricanes to occur? and (2) Does global warming cause hurricanes to be more intense?

---

*We'd like to take this opportunity to lobby the World Meteorological Organization to name hurricanes Seth and Doug, though on second thought perhaps it's better not to have our names associated with catastrophic events that ruin—or even take—people's lives and livelihoods. Actually, in the Atlantic and Eastern North Pacific (the locations of the vast majority of tropical cyclones), there are predetermined lists of names that are rotated on a six-year basis. The names are recycled over and over with the exception of when there is a particularly deadly/costly storm (like Sandy or Katrina)—then that name is retired and replaced with a new one . . . so there's a possibility for a Hurricane Seth or Doug yet; we'd just need Sebastien, Sally, Sam, Shary, Sean, or Sara, or Dorian, Dolly, Danny, Danielle, Don, or Debby to be a whopper.

The scientific community doesn't have a clear answer to the first question yet. If you look at the recorded number of tropical storms and hurricanes in the North Atlantic over the past century, you see a clear upward trend. It could be tempting to attribute this to global warming, but the upswing may also be a result of better monitoring in recent years with aircraft, radar, and satellites. Climate models don't help here, either, because some models say yes, global warming is likely to cause more hurricanes, while others (in fact, quite a few) say no. So, the jury is still out on the question of hurricane frequency.

The jury is basically in on the second question. There is a growing consensus, backed by strong evidence, that global warming is making hurricanes more powerful. In order to determine if there has been a trend in hurricane strength, we need a way to quantify the power of a storm. Since the 1970s, the Saffir-Simpson Scale has been the standard for classifying hurricanes. The scale, created by Herbert Saffir and Robert Simpson, is straightforward: It simply uses the storm's maximum sustained wind speed—that's it. There are five ratings, or categories, on the scale. For example, a Category 4 hurricane has sustained winds of 58–70 meters per second (130–156 mph). The Saffir-Simpson Scale's simplicity is also its limitation. Because it only looks at wind speed, it doesn't account for the size of a storm's wind field or its capacity to cause coastal inundation when it makes landfall.

October 2012's Hurricane Sandy, for example, briefly reached a maximum intensity of Category 3 and weakened to a Category 1 storm closer to landfall. The low classification led some to underestimate its destructive potential—a costly error in this case. Had the residents of New Jersey and New York known ahead of time that Sandy was far more powerful than even Hurricane Katrina, they might have prepared for the storm differently.*

---

*It's worth pointing out that Sandy was actually an oddball phenomenon—a mix between a hurricane and something called a "midlatitude synoptic storm" (leave it to meteorologists to come up with such a boring term for something so powerful). This sort of hybrid phenomenon is in a different category from your typical hurricanes and may be impacted by global warming in different ways.

Among the more useful hurricane metrics for our purposes is the Power Dissipation Index, which is a measure of the total dissipation of power, integrated over the lifetime of the storm.* When this index is tracked from the 1970s until today (that is, during the period in which anthropogenic global warming has been most prevalent), there is a clear increase in the power of hurricanes. This trend is the result of both longer-lasting storms and more-intense storms.

In addition to letting us chart the power of storms, the index is useful because we can see how the trends it reveals compare to the trends we see when we measure sea-surface temperature. The temperature at the surface of the ocean goes up and down over short periods, but it has a clear upward trend in the long term because of global warming. Both the short-term and long-term changes in sea-surface temperature match changes in the hurricane Power Dissipation Index, which means that as the planet—including the seas—warms up further in the future, we can expect even more powerful hurricanes than we've experienced so far.

There is another factor related to climate change that is driving the destructive power of hurricanes and other coastal storms: sea-level rise. Storms aided by sea-level rise have increased the risk of storm-surge damage, erosion, and flooding for coastal communities. Over the past century, global sea level has risen by approximately eight inches—and the rate of sea-level rise is accelerating. Sea level is projected to rise by another one to six feet this century!

Exactly how much higher it goes will depend on how much more greenhouse gas we add to the atmosphere. Higher carbon dioxide emission scenarios would lead to more warming and an associated increase in ice melting, and would therefore be expected to lead to sea-level rise toward the upper end of the projected range. The stakes are grave. Nearly five million Americans live within four

---

*Others include the Accumulated Cyclone Energy Index, the Integrated Kinetic Energy Index, and the Hurricane Severity Index.

feet of the local high-tide level. Broad regions of densely populated South and Southeast Asia are also extremely low-lying land.

Global warming is likely causing stronger storms, which are having a greater impact on our coasts because of elevated seas, but there is even more bad news. Climate disruption and other human influence on ecosystems often increases their vulnerability to damage from extreme weather events. And this is coupled, tragically, with reduction in their natural capacity to temper the impact. Coral reefs, mangrove forests, barrier island chains, and salt marshes protect coastal ecosystems and infrastructure against storm surges. Floodplain wetlands absorb floodwaters. Anthropogenic effects, in many cases connected to global warming, are decimating these protective landscapes. Not only are stronger storms on the horizon, but we are more defenseless against them than ever.

# Climate Change Isn't Happening

Feels pretty cold . . . where's your global warming?

The planet isn't getting warmer

Glaciers are growing, Antarctica is gaining ice

Climate is too complex to model or predict

# 6

# Feels pretty cold . . . where's your global warming?

**SO, WE GET** a cold spell and suddenly global warming is not a problem? Seemingly every time it gets a little nippy outside (can you say "polar vortex"?) or there's a particularly impressive winter storm (can you say "Snowmageddon"?) we hear from skeptics that climate change isn't happening. Understandably, it can be difficult to concern yourself with global warming when you're busy shoveling your car out of a gargantuan snowdrift, but as we'll explain, not only is this sort of thing *not* a sign that global warming isn't happening; in some cases it may itself even be an effect of global warming.

One of the common errors of climate skeptics is a confusion of weather and climate; another is a confusion of local effects and global trends. Both are at play here. As we introduced in the Prologue, the regional climate is simply the thirty-year average of the weather where you live. Global climate, on the other hand, is the average of all the regional climates of the world. Climate normals (thirty-year averages) were established in accordance with the recommendation of the World Meteorological Organization (WMO) starting with the period from 1931 to 1960.

Meteorologists and climatologists use normals to provide context for recent climate conditions. For example, when you watch the TV weather, the meteorologist tells you not only the high and low temperatures (among other weather variables) for that day, but also the "normal" high and low for reference. These are those thirty-year averages for temperature where you live. Since these are averages, the daily temperature is rarely exactly the same as that thirty-year normal. This is because the Earth's weather is dynamic and does not behave exactly the same way each day. Even so, the regional climate remains the same because the numbers of days above and below the average temperature balance out over the course of seasons, years, and thirty-year timespans—if the Earth's climate isn't changing. The same is true for the other weather variables like humidity, wind speed and direction, pressure, precipitation, cloudiness, and so forth.

You might experience a string of especially warm days in a row (days above the regional climatic normal) where you live, but things have a way of averaging out over time. In other words, if you collect enough data, the true average temperature will be revealed. Statistically speaking, it should be just as likely that you will get a string of cold days in a row. Since we are all tuned in to "global warming," it is a natural reaction (especially for Brad) to think "Aha, it's been colder than normal all week . . . so where's your global warming?" In fact, we would have to track differences between the daily highs and lows and the regional climatic normal highs and lows for those days for thirty years before we could say something definitive.* Good thing organizations like the National Weather Service keep those records! But of course the long-term averages are not actually constant—that's what global warming means.

What is interesting is that the Earth's global average temperature has increased, but only by about 0.8 degrees Celsius (1.4 degrees Fahrenheit), since the early twentieth century, with about two-thirds of the increase occurring since 1980. This doesn't seem like much, but it is certainly enough to melt all the planet's ice over time.† Importantly, this is the change in the global climate average,

---

*And sometimes temperatures in a certain region can be colder (or warmer) than normal for months on end, as was the case with the early 2014 North American cold wave driven by an unusual polar vortex. Interestingly, weather events of this nature are expected to become more common as a result of climate disruption (see Chapter 5).

†This relatively modest amount of warming that we've already experienced would take centuries to melt all of the surface ice on Earth, but as warming becomes stronger, the pace of melting will accelerate rapidly.

not a regional climate average increase. Regional climate averages are not the same everywhere. This number can be much larger (or smaller) where you live. Let's look at some examples.

In commemoration of Earth Day, 2013, Climate Central (a United States–based non-profit organization of scientists and journalists who do research and report about our changing climate and its impact) created an interactive graphic that shows a state-by-state analysis of average temperature trends since the first Earth Day, in 1970. Overall, temperatures in the continental United States have been rising gradually since the early twentieth century. That trend parallels an increase in global temperatures. The increase was so subtle in the early years that neither scientists nor environmentalists took much notice, and in the United States, some states had shown no warming at all. Starting in 1970, the increase in temperature began to accelerate. Every state's annual average temperature has risen significantly since then, at an average rate of 0.435 degrees Fahrenheit per decade—about triple the national average of 0.127 degrees Fahrenheit over the years 1910–2012.

Those are the averages, but state-by-state differences remain. The fastest-warming states—Arizona, Michigan, New Jersey, and Minnesota—have warmed by about 3 degrees Fahrenheit over the past forty-three years, or about 0.7 degrees Fahrenheit per decade; about twice as fast as Earth as a whole. The slowest-warming states—Washington, Oregon, Alabama, and Georgia, with an increase of about 0.35 degrees Fahrenheit per decade—are more or less keeping pace with the global average. On a larger scale, regions such as the Pacific Coast and the Southeast are warming more slowly, while the Northeast and north-central states are warming faster. Of the forty-eight contiguous states, thirty have warmed by more than 2 degrees Fahrenheit since 1970, and seventeen have warmed by 2.5 degrees Fahrenheit or more.

Remember that all this long-term warming doesn't mean, however, that there won't be cold days—even really, really cold days—as we can attest to, experiencing Chicago winters! It's natural to remember those really crazy weather days more easily than subtle long-term trends. Climate scientists usually focus on the average temperature trends, and Chapter 7 will cover the evidence proving that the long-term trends do indeed clearly show warming on a global scale, but let's take a look at these crazy weather days—they have something to show us about climate change, too.

As we said, meteorologists give us the forecast temperatures and normal high and low temperatures for a given day, but a third thing they also often talk about is the *record* highs and lows for that day. These are the crazy days, like August 28, 1986, when it was only a chilly 42 degrees Fahrenheit (5.6 degrees Celsius) in Chicago—or, conversely, March 21, 2012, when it was a whopping 87 degrees Fahrenheit (30.6 degrees Celsius). Over time, it gets increasingly harder to break these temperature records because the data set keeps growing. To understand this, think about the very first year in which temperatures were recorded; every day that year would set the record high and low temperatures for that date. The next year, many days would break those records, but probably not every day. The third year, still fewer days would be records, and so on. If you plot the total number of records over the entire United States vs. time, the number of records does in fact drop as expected. But if you dig some more into these data, you can find another trend— one that indicates global warming is indeed happening.

Separating out the records into record high temperatures and record low temperatures is informative. If the climate were stable, you would expect the numbers of record highs vs. record lows to be approximately the same over time. This is not the case! In the '70s, the ratio of highs to lows was 0.78:1. In the '80s it increased to 1.14:1, followed by 1.36:1 in the '90s and 2.04:1 in the 2000s. So since the 1970s,* we have been experiencing progressively more record highs than lows. That's a very clear trend, and it has occurred over a sufficiently long period of time to be significant.

---

*To learn why this trend started in the 1970s (and not earlier), check out Chapter 7.

We are now seeing about twice as many record highs as record lows in the United States. Obviously, while you can still have days that break records for being cold, overall, things are getting substantially warmer.

So we've explained how a few cold days—even record cold days—don't mean that global warming has stopped, but just to make sure that we put this line of skeptic arguments to rest, let's also explore somewhat longer cold spells.

One cold snap that captured the attention of skeptics took place from December 2009 through February 2010. Most of Europe, Western Asia, and North America experienced an unusually cold and snowy period during this span. One particularly vocal skeptic, Senator Jim Inhofe (R-OK), and his family even went so far as to build an igloo near the White House in Washington, DC, with a cardboard sign mocking the "global warming scam," as if to say, "Obviously there is no global warming, since we've got enough snow to build an igloo."

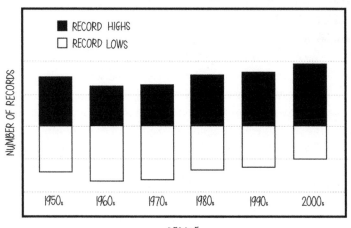

Ratio of record high and low temperatures in the US by decade

Aside from the fact that, as we've explained, a stretch of cold weather is perfectly consistent with a warming planet, you can gain additional perspective on this specific cold spell by zooming out a bit. Eurasia and North America were unusually cold, but at the same time, large portions of the Arctic (Greenland, Canada, and the Arctic Ocean) were unusually warm, as were North Africa, the Mediterranean, and Southwest Asia. The big swings in temperature by region were the result of an unusually strong Arctic Oscillation—the strongest occurrence in at least half a century. The Arctic Oscillation is a cyclic shifting of sea-level air pressure in the Arctic. Changes in the air pressure in the Arctic can influence weather thousands of miles away by pushing the jet stream around. Those big snowstorms in 2009 and 2010 may have made for good PR for climate skeptics, but they in no way proved that global warming is not happening.

In fact, some climate scientists argue that extreme precipitation events like the February 2010 Snowmageddon (or do you prefer Snowpocalypse? Or Snowzilla?) are an *expected outcome* of a hotter Earth. That month witnessed two "hundred-year" snowstorms separated by just a few days. We've spent a good deal of time on the connection between climate disruption and extreme weather in general in Chapter 5, but given skeptics' attachment to this particular series of storms as supposed evidence against global warming, we'll revisit the connection a bit here. Global warming increases evaporation from the oceans. The correlation between this extra water vapor and atmospheric humidity levels is complex and is still an active area of study by climate scientists, but there is a logical connection between increased evaporation and increased precipitation. It is a water cycle, after all; water that goes up generally comes back down. Heavy precipitation, which during a cold snap will fall as snow, is therefore fully consistent with global warming.

Not only are additional monster snowfalls possibly correlated with climate change, but their location is also changing. For a snowstorm to occur, air temperatures generally need to be in the range between negative-ten and zero degrees Celsius (fourteen to thirty-two degrees Fahrenheit). Higher temperatures result in other types of precipitation, and lower temperatures are not conducive to heavy snowfall. This is because at lower temperatures there is lower moisture content and insufficient "lifting," the dynamic process of air flow by which snowflakes develop. This means that, as the planet warms up, the regions where snowstorms are likely will shift along with those regional temperature changes.

Observations back up this phenomenon. On average, there have been fewer snowstorms in regions such as the lower Midwest and the South in the United States, and more snowstorms in regions such as the upper Midwest and the Northeast. (The overall national trend in the United States is also for more snowstorms.)

One last point to make about snow is that there is a distinction between snowfall and snowmelt. While snowfall may increase (at least for now and only in certain regions) because of global warming, snow cover is expected to decrease earlier in the year as a result of increased melting. The Rutgers University Global Snow Lab (cool place to work!) compiles data on snow cover in the Northern Hemisphere. For our discussion, a particularly telling piece of these data is the springtime snow extent, with data going back to the late 1960s, because the spring is obviously when snow usually melts. There is an indisputable trend of decreasing snow cover found in these data, with an average loss of over 30,000 square kilometers of snow per year. That's more than one and a half times the area of the entire state of New Jersey!

It turns out that the snowmelt is more than just a sign of global warming (see Chapter 7)—it's an active participant. Bright, reflective snow sends sunlight back into space. When it melts, leaving behind darker, exposed ground, more heat from the sun is absorbed, which drives further warming. Frighteningly, this feedback process is taking place even faster than has been projected by most climate models. So, Brad, there's our global warming.

# 7

# The planet isn't getting warmer

**YOU MIGHT BE** wondering why we'd wait until Chapter 7 to get to what is probably the most common argument put forward by skeptics like Brad: the assertion that climate change isn't even happening. Yup, you might think someone would be crazy to suggest not that there are alternative explanations for global warming but that the globe isn't even warming in the first place. (If you think this one is crazy, you're definitely going to get a kick out of some of the following chapters.) It certainly would make things an awful lot easier for all of us if this idea were true. We'd all be able to take a big collective breath, fire up the Hummer for a drive to buy some imported beef, and stop fretting about those adorable polar bear cubs.

But now that we've covered some of the basic concepts, we can really dig into this one. Here's the gist of Brad's argument: He says global warming in recent years has paused, or maybe even reversed into global cooling. As with most climate-change myths, this one draws on some actual (though carefully cherry-picked) data to make the case, but seems to willfully ignore the overall set of data that contradicts the claim. Although subtleties of several of the individual topics we'll consider are still being studied, discussed, and debated by the scientific community, the only phenomenon that can explain them all is anthropogenic climate change. It is the full body of evidence that provides such a convincing case for our impact on the climate.

One really important point to make here is that the air and the land are only a small piece of the Earth's climate. It is called *global* warming for a reason. Admittedly, we humans consider these particular parts of the climate pretty important, since we spend most of our lives on land and breathe the air, but to understand climate, you have to account for the biggest home for additional heat on the planet: the oceans. Global warming makes the atmosphere hotter, which is generally front and center in conversations about climate change, but it also heats up the land, melts ice, and warms the oceans. It's this last part that is most often overlooked or misunderstood. Yet the heat content of the oceans dwarfs that of the land, atmosphere, and ice put together. This means that even small exchanges of heat between the air and the oceans can strongly influence the temperatures measured at or near the surface of the Earth.

Such exchanges explain many of the relatively short-term variations in temperature that we experience as the land-dwelling, air-breathing creatures that we are. El Niño and La Niña are different stages in a cyclical pattern of climate turbulence known by meteorologists as the Southern Oscillation. First noticed by sixteenth-century fisherman on the Pacific coast of South America, these phenomena were not scientifically documented until the 1920s, when scientists noticed periodic widespread changes in weather every three to seven years in concert with air-pressure changes in the eastern Pacific. (It took another few decades for scientists to figure out the connection between these air-pressure changes and ocean temperature changes.) Since the 1970s, though, El Niño and La Niña have become more frequent and intense. Although scientists can identify these events and their influence on global weather patterns, they still do not know for sure what causes them. The growing consensus is that the Southern Oscillation is driven by the Earth's climate. The oscillation between El Niño (warmer than normal) and La Niña (cooler than normal) is simply an overshoot problem: It takes six months or more to get an El Niño cycle going in weather patterns around the world. Then it shuts down. An analogy to these overshoots in climate turbulence would be the way your thermostat that controls your furnace or air conditioner clicks on at some set temperature, but it takes a while for the heating or cooling to kick in so the temperature keeps moving for a while longer.

Some scientists think that the increased intensity and frequency—now every two to three years—of El Niño and La Niña events in recent decades is due to warmer ocean temperatures caused by global warming. In a 1998 report, scientists from the US National Oceanic and Atmospheric Administration (NOAA) explained how higher global temperatures increase evaporation from land, thereby adding moisture to the air, which in turn intensifies storms and floods associated with El Niño.

Another possible explanation is that the Southern Oscillation may be like a pressure-release valve for the tropics. With global warming driving temperatures higher, ocean currents and weather systems might not be able to release all the extra heat pumped into the tropical seas. El Niño, then, helps to expel the excess heat. Either way, increased carbon dioxide from the use of fossil fuels is the primary source of the additional heat to which nature is responding. For example, 1998 was a strong El Niño year, leading to heat moving from the ocean to the atmosphere and one of the hottest years on record. La Niña cycles have the reverse effect, and the combination of these two phenomena is one of the causes of the year-to-year fluctuations in surface temperature.

Another common factor that affects temperature on a short timescale is volcanic eruptions, which blast sulfate aerosols (and carbon dioxide) into the atmosphere. These aerosols reflect some sunlight back into space, leading to temporary cooling (whereas the carbon dioxide leads to longer-term warming). Because of gravity, the particles eventually fall back to Earth, but the carbon dioxide stays in the air much longer.

Natural cycles of solar activity are still another short-term influence on climate. This topic warrants its own discussion in a later chapter (11), because climate-change skeptics like to pull it out of their hats as an alternative explanation for the increase in global warming that we observe.

Even taking into account these causes of short-term cooling, the overall heat content of the system as a whole just keeps going up. In fact, our climate is accumulating heat at an average rate of nearly 200,000 gigawatts (take that and stick it in your flux-capacitor time machine, Doc Brown!*), almost all of which is being absorbed by the oceans. Think of it this way: The heat increase is essentially the same as taking the power from more than one hundred thousand utility-scale power plants and dumping it straight into the ocean. The take-home message from this is that when people try to make claims about climate change (or the lack thereof) based only on surface-temperature data from a short period of time, you should immediately be distrustful of their conclusions.

OK, back to Brad and his claim that global warming has paused or reversed. He not only says that the upward trend in surface temperature has leveled off in recent years, but that the oceans even cooled in the middle of the past decade. Turns out, Brad is right! Well, sort of. Let's start with the oceans, which account for most of the Earth's heat content. He's right that there are data from a worldwide network of thousands of floats called the Argo buoys that, among other things, measure temperature near the surface of the oceans (upper-ocean temperature), and seem to show that the oceans have been cooling since 2003.

Short-term variations can be substantial, but it's the long-term trend that is the marker for climate change. For example, you can see a correspondence between the volcanic eruption of Mount Agung in Bali, Indonesia, in the early 1960s and a decrease in the amount of heat in the world's oceans over several years. There is no question, though, that, long-term, upper-ocean temperature is increasing. (Moreover, much of the heat is stored at greater depths in the ocean than are measured by these studies.)

Let's give the claim about the recent ocean cooling a closer look. You might think that measuring ocean temperature would be

---

*For the *Back to the Future* fans out there.

a pretty straightforward exercise. Just stick a thermometer in the water, right? Unfortunately, it's more complicated than that. Before the Argo network was established, ocean temperatures were measured with things called expendable bathythermographs, or XBTs. Careful analysis shows that XBTs have a warming bias in their measurement; that is, they measure the temperature as being higher than it actually is. It turns out that the Argo buoys have an opposite, or cooling, bias associated with their pressure sensors. Guess what year the Argo buoys were deployed? Yeah, it was 2003. Over the next four years, more and more buoys were used, enabling the network to take more thorough samples of ocean temperature. When you put these two biases together, you have the making of what appears to be a cooling phenomenon. This is because the XBTs made the temperatures seem higher than they were (warming bias) earlier in the temperature record, whereas the Argo buoys made the temperatures seem cooler than they were (cooling bias) in the more recent temperature record. Over time, then, the temperatures have appeared to cool even though they haven't. Fortunately, knowing about these biases allows researchers to correct for them. Researchers are still deciding how exactly to put a precise number on the size of the temperature-bias phenomenon, but their preliminary conclusion is that there has been a slight warming since 2003. Regardless, the long-term trend, which is the one that matters, is most definitely one of warming.

Although long-term warming of the ocean is a strong indicator of global warming, it is far from the only one. In fact, NOAA released a report in 2010 that covers a broad spectrum of climate indicators. One of these is land-surface temperature, which is the target of seemingly endless attacks by folks like Brad. But before we explain why these attacks do not jibe with the facts, let's go over the many other indicators that *all* point to a warming planet.

Not only does land-surface temperature show a long-term trend of getting warmer, so too do sea-surface temperature and marine air temperature (the air over the oceans). Data going back well over a century show an unambiguous increase in both. Check and check. The lowest part of the atmosphere is called the troposphere. There are good temperature data on the troposphere going back to about 1960. Warming there, too. Check. If the world is warming up, humidity should be rising as more water evaporates into the atmosphere. So what do the humidity data* from 1970 to today show? OK, you know where we're going with this. Check. Sea level? We've got continuous measurements extending all the way back to the 1800s from tide gauges (essentially tubes with sensors that measure the height of the water level). Not only is sea level rising over the long term, but this rise is also accelerating. Check. Another set of indicators has to do with melting ice or snow. Glaciers have been steadily losing mass since about the middle of the twentieth century. The same goes for snow cover in the Northern Hemisphere and for the Arctic sea ice. Check, check, and check. Or is it checkmate at this point?

Actually, there's more. Spring is coming earlier. Jet streams—those fast-flowing air currents you see on weather maps that move west to east and influence air travel and weather—are moving closer to the poles. The Arctic permafrost (frozen soil) is thawing to greater depths each summer and degrading overall. Biological organisms are reacting to all this upheaval. Earlier seasonal changes mean that creatures such as butterflies and frogs emerge from their cocoons and ponds earlier than they did in the past. The distribution of animals and plants is also changing, with trees, birds, insects, fish, and many other living things all shifting closer to the poles.

---

*Humidity is a trickier thing to measure than you might think. The absolute level of humidity (amount of water vapor per unit volume) is generally calculated from measurements of temperature, pressure, and relative humidity (ratio of how much water vapor is in the air relative to the total amount that the air could hold). Absolute humidity varies a lot over a day as well as longer timescales, so data can jump around quite a bit. Still, the evidence today does indeed indicate a clear trend of increasing humidity levels in the past few decades.

# LINES OF EVIDENCE FOR GLOBAL WARMING

OCEAN HEAT
CONTENT RISING

SEA SURFACE
TEMPERATURE RISING

MARINE AIR
TEMPERATURE RISING

WARMING
TROPOSPHERE

HUMIDITY RISING

SEA LEVEL RISING

SHRINKING GLACIERS

SNOW COVER DECREASING
IN THE NORTHERN HEMISPHERE

ARCTIC SEA
ICE MELTING

LAND SURFACE AIR
TEMPERATURE RISING

All right. So we've got nine independent lines of evidence that the Earth is getting hotter in the long term, strengthened by a set of secondary indicators. The temperature of the air near the surface of the land—the temperature we all experience on a daily basis—makes ten. Despite herculean efforts by Brad to discredit this particular indicator, it, too, reliably supports the case for global warming. We will spend the rest of this chapter addressing Brad's various criticisms of this indicator.

Brad's ready for you this time. He says, "How can the world be getting hotter if 1934 was the hottest year on record?" If you're as sharp as we think you are, you're ready with a quick response: Picking a single year out of the temperature record doesn't have much to do with long-term climate change, because short-term variations push things up and down all the time. Well done! We've got plenty more for you to keep in your pocket, though, in case Brad is as stubborn as we both know he tends to be. The year 1934 was indeed a hot one in the United States, but not the hottest. Mind you, it was surely an uncomfortable one for all those folks without air-conditioning. However, 1998, 2006, and 2012 were all hotter years in the United States. But a more important point is that the United States represents only a tiny fraction of the surface area of the Earth. When you look at the global temperature record, 1934 wasn't so hot after all. So Brad has committed a double cherry-picking penalty. (That's four minutes in the penalty box, Brad.) He made the mistake not only of picking a single year but also of looking only at a relatively local phenomenon. It bears repeating: We're talking about *long-term global* warming. Globally speaking, the ten hottest years on record were all since 1998.

While steaming away in the penalty box, Brad comes back with the point that skeptics recently identified an error in the way corrections to temperature data (to accommodate biases in the measurements like the ones we mentioned with the XBTs and Argo buoys) were handled by NASA and NOAA. More precisely, researchers there were inadvertently using uncorrected data starting with the year 2000 (known in the skeptic community as the "Y2K glitch"). Kudos to the skeptics here. You've got to give them credit for keeping everyone on their toes. The corrections in question affected only the continental United States, and these temperature data from NASA and NOAA are a very small part of the global temperature record, but regardless, let's take a look at what happens to the data once the adjustments to correct for measurement biases are properly applied. As with all trends in climate change, it's important to view things over a sufficiently long time, say from 1975 to today*—the period over which the influence of our greenhouse-gas emissions on global warming has dominated. And, of course, we have to view the global average, not just the US data. Before the data correction, the warming trend over this period is 0.185 degrees centigrade per decade. (Brad is on the edge of his seat, ready to throw an "I-told-you-so" at you.) After correcting the data, the warming trend drops to—drumroll please—0.185 degrees centigrade per decade. Seriously? Yeah. The correction changes the trend by less than one thousandth of one degree. As with the Y2K millennium computer bug, lots of hype with nothing to worry about in the end. Except for the fact that it means climate change is real.

Brad's out of the penalty box now and ready with another one: "So you claim that only long-term trends in temperature are meaningful from a climate-change perspective. We were burning plenty of coal back in the middle of the twentieth century. How come the temperature went down during that period?" Before you roll your eyes, we have to admit he has a point there. The period in question is about 1940–1975, which should be long enough to be considered relevant for climate change. We've mentioned a few natural causes that can produce short-term cooling, but this cooling trend isn't El Niño/La Niña or cycles of the sun or volcanoes at work. The record definitely shows growing emissions of carbon dioxide during that time. Is he wrong that

---

*We're using the year 2012 for "today" since the data are complete through 2012 as of this writing.

global temperatures cooled then? Nope. He's spot on. Now, we're not talking about a huge drop here—only about 0.1 degree centigrade of cooling. Nonetheless, it was cooling, and this was a departure from the periods both before and after this span (1900–1940 and 1975–today). So what gives? This turns out to be a fascinating story, and it brings back those sulfate aerosols we mentioned earlier in the chapter. Only this time, they aren't coming from volcanoes.

After the Second World War, there was a period of economic expansion, which translated into industrialization on a massive scale. That growth lasted until the early 1970s, when the oil crisis, a stock-market crash, and other forces slowed things down. This industrialization substantially increased emissions not just of greenhouse gases but also of many other molecular troublemakers. We know all about carbon dioxide and its effect on global temperature. Burning fossil fuels releases all sorts of other stuff, too. One of these other emissions is sulfur dioxide. When sulfur dioxide mixes into the atmosphere, it forms tiny droplets of sulfate solution, or sulfuric acid. These tiny droplets, along with very small solid particles of sulfates, are known as sulfate aerosols. These little buggers can affect climate in several ways, ranging from direct scattering or absorption of sunlight to influencing clouds and airplane contrails (affecting their number, size, and duration as well as their optical properties). Scientists are still working to understand these complex interactions, but the general consensus is that sulfate aerosols have a net cooling effect on climate owing to the scattering of sunlight back into space. So greenhouse gases are just one of a collection of things that affect climate. Some things push toward warming, some toward cooling, and it is the balance of these various forces that results in the observed long-term trends.

Even though the planet cooled a tad during this thirty-five-year stretch, we can still find strong evidence of global warming happening at the same time. Sounds like a contradiction? Well, let's work through it. Here we want to disentangle many different factors, and digging deeper into the temperature record can help. Rather than just looking at the overall average temperature, let's compare the daytime maximum and nighttime minimum temperatures. During this period in the middle of the twentieth century, the average maximum daytime temperature fell. This makes sense, because of the scattering of sunlight by sulfate aerosols.

But at night there is no sunlight to scatter, so what about the nighttime minimum temperatures? They were rising! In other words, the aerosols were causing global dimming that reduced the temperature in the day, but greenhouse gases such as carbon dioxide were warming things up underneath it all.

Enter the environmental movement, stage left. Those sulfate aerosols have other effects, too, such as causing acid rain. And there are plenty of other nasty compounds in emissions from fossil-fuel power plants (toxic heavy metals, particulates, ozone . . . the list goes on). Eventually, the environmentalists were able to raise public awareness of these issues and establish sufficient political momentum for landmark legislation to be passed in the 1970s, including the Clean Air Act, the National Environmental Policy Act, and the Clean Water Act. One of the practical outcomes of these laws was the implementation of measures to control pollutions from fossil-fuel power plants. This change led to dramatic and rapid halts to the growth of a swath of emissions (though, notably, not of carbon dioxide*). The air became cleaner (which no doubt was appreciated by more than just Dr. Seuss's swomee swans). The concentration of sulfate aerosols decreased precipitously, along with their cooling influence on the climate. Sure enough, starting around 1975, land-surface temperature across the world resumed its relentless rise, a trend that continues to this day. Before you go getting ideas about ripping out all those pollution-control measures in an effort to combat climate change, keep in mind that greenhouse warming was still going on all that

---

*Much more recently, the US Supreme Court ruled in *Massachusetts v. EPA* (2007) that greenhouse gases do in fact count as air pollutants under the Clean Air Act and are therefore appropriate to be regulated by the federal government.

time. It was just masked by the aerosols, which themselves caused plenty of infrastructural and environmental damage.

As relentless as ever, Brad has one more thought to throw at you to support this whole the-planet-isn't-actually-warming myth. "Maybe these land-surface temperature measurements show a long-term increase, but aren't a bunch of these thermometers sitting in urban areas? And don't urban areas have a 'heat-island effect,' which will artificially make it look like the average global temperature is higher than it really is?"

Well, the urban heat-island effect is a real thing. Many urban areas tend to have warmer air over them than do rural areas. This is a different type of warming, and a highly localized one at that, not to be confused with global warming. Materials such as concrete and asphalt retain more heat than do natural surfaces; waste heat is also generated by our use of energy in cars, air-conditioning, and so on. So more heat is going to be generated in places where there are more of us—in other words, in urban areas. Because urban development has been expanding, argues the skeptic, the observed temperature rise can be attributed to the urban heat-island effect rather than to the greenhouse effect.

Climate scientists, being the meticulous people they are, go to great lengths to make sure that the urban heat-island effect does not skew temperature trends in their analyses and make average global temperatures appear higher than they really are. NASA and NOAA teams scrutinize urban long-term trends and neighboring rural long-term trends, and whenever they identify a statistically significant discrepancy, they adjust the urban trend to match the rural trend, operating on the (very reasonable) assumption that the rural areas will not be impacted by the urban heat-island effect and are therefore more reliable measurements of the local temperature. In most cases, it turns out, the urban heat-island effect is actually really small and does not require any correction to the data. (In over one-third of the cases, urban trends were actually cooler than those in their neighboring rural areas!)

So if the urban heat-island effect is real, how come it isn't having a big influence on the temperature data, or more important, on the temperature *trends*? Part of the explanation is that meteorological stations in cities are often located in cooler areas, such as parks, as opposed to industrial areas, where the heat-island effect is the strongest. In other words, researchers have sited these weather stations in places likely to minimize the urban heat-island effect. As for the trends, they are due to climate change, not urban development, so of course there is not a big urban heat-island effect on them. The take-home point here is that the data that reveal long-term climate change already account for any urban heat-island effect, so this myth is busted.

Oh, and just to keep Brad from popping up with some snarky reply, you can add that if you look at where most of the warming on the planet has occurred, you find that it's in the polar regions. Not many people there.

# 8

# Glaciers are growing, Antarctica is gaining ice

**AS WE'VE MENTIONED** a number of times in this book, the effects of climate change often show up first and/or most strongly in the polar regions, so it should come as no surprise that skeptics go after these effects in their attempts to show that the planet isn't warming after all. However, like the critters in a sick Whac-A-Mole game, the facts just keep popping back up despite their best efforts.

The interplay of ice and a warming planet is not as simple as it might seem, but melting of ice is indeed a classic signature of climate disruption. If you're going to hunt around for this effect, it makes sense to start with places where there's lots of ice, and there's certainly lots of ice both in Antarctica and in the Arctic region (including Greenland). There's also a lot of it locked up in glaciers at high altitudes around the world.

Brad latches on to reports that sound pretty damning to the concept of anthropogenic climate change. He says Antarctica is *gaining* ice and that glaciers are *growing*. As is so often the case, what we have here is a glaring incident of cherry-picking data to suit one's argument rather than looking at the big picture. In Brad's defense, the subject of ice changes is a pretty complicated one. Ice comes in many flavors (and we don't mean lemon vs. strawberry at the ballpark), and confusing them can lead to serious misconceptions like those held by Brad. We'll start with Antarctica; it's got an interesting story to tell.

The two main flavors of ice are sea ice and land ice, and knowing this distinction will go a long way toward understanding Brad's mistake. Sea ice is ice formed from salt water and is floating in the ocean, the coverage of which cycles substantially over the course of a year. Because the water locked up in sea ice is already displacing seawater, melting or formation of sea ice does not have much effect on sea level. This is much like how when the ice in your drink melts, it doesn't cause the liquid to spill over the edge of your glass. Land ice, in contrast, is ice that has accumulated over many, many years through snowfall. That water used to be in the oceans before it evaporated and fell as precipitation, so storing it as land ice means the sea level is lower than it would otherwise be, and, obviously, if it melts and returns to the ocean, sea level will then rise (see Chapter 3 for some consequences of sea-level rise).

It seems apparent that if the Earth is warming, ice should be melting in the long term. Despite this, it is true, as Brad points out, that Antarctic ice has been growing. But there's a big "but" to attach to this fact. It is Antarctic *sea* ice that has been growing. Satellite data going back to the 1970s show a slow but clear trend of an increasing area covered by sea ice (known as the "sea ice extent") in that region. This really is a bit of a puzzle, because not only are the oceans warming (see Chapter 7), but the Southern Ocean is actually warming even faster than the others—about 0.17 degrees Celsius (0.31 degrees Fahrenheit) per decade. Why the heck would a warmer ocean have more ice?

Bear with us on this one, as it takes a few steps to explain. Remember that ozone hole that had everyone so freaked out back in the 1980s?* Well, it's still up there over the South Pole, and it causes localized atmospheric cooling. This is because ozone absorbs UV radiation and as a result the atmosphere heats up—so less ozone means less heating. When the stratosphere cools, the cyclonic winds that circle Antarctica get stronger and push the sea ice around, and this exposes more open water† to freezing and, voilà, more sea ice

---

*For those of you too young to remember it, be glad you weren't alive when "Don't Worry, Be Happy" was inundating the airwaves and parachute pants were all the rage.

†These areas of open water are called "polynyas" . . . fun Scrabble word to keep in your pocket.

forms. This particular mechanism is still being debated by climate modelers, but there are other mechanisms at play.

There's a second reason for the recent increase in sea ice: ocean circulation. Think of the Southern Ocean like hot chocolate with cool whipped cream on top: a layer of colder water near the surface and warmer water underneath. Normally, the warmer water would rise up to the top because it's less dense than the cold water. But things aren't normal, thanks to our disrupting the climate. Global warming means warmer air temperatures, which in turn means more precipitation. All that extra fresh water falling as rain or snow makes the surface water less salty, and fresher water is less dense, so even though the water is cold, it is less dense than the saltier warm water underneath. That translates to less ocean circulation, less heat moving upward, and less sea ice melting.

GROWTH OF ANTARCTIC SEA ICE

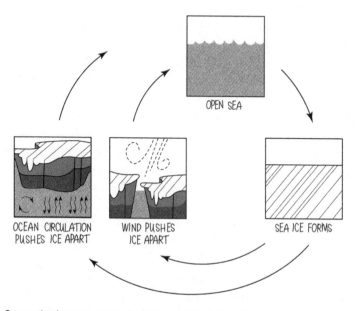

OPEN SEA

SEA ICE FORMS

WIND PUSHES ICE APART

OCEAN CIRCULATION PUSHES ICE APART

**Connection between ocean circulation and formation of sea ice**

A third, but less significant, reason for the increase in Antarctic *sea* ice is that Antarctic *land* ice is melting and thereby freshening the surface water further still. And, again, fresher water decreases ocean circulation, which in turn decreases the melting of sea ice. So, although it may seem counterintuitive at first, a warming planet actually can lead to more sea ice—assuming the conditions are right. Note that, in the long term, this trend is expected to eventually reverse as global temperatures continue to creep even higher, because the balance will tip in favor of melting over the sea-ice growth mechanisms we just described. Also, the Antarctic sea ice that we've been talking about does not play nearly as large a role in future global warming as does the Arctic sea ice, which sticks around throughout the year. Since the Southern Ocean has essentially been ice-free in the summer for many years, the sun's radiant energy is already being absorbed rather than reflected back into space like it is in the Arctic summer. Loss of Arctic sea ice is a major global warming feedback concern; the Antarctic sea ice, not so much.

Let's shift to the other flavor of ice, land ice. This is the ice that's important for sea-level rise. It's important to state up front that measuring the mass of land ice is a non-trivial task. If it were just about measuring the area covered by ice, satellite imagery could do a bang-up job. But what's important is the volume of the ice; that is, not only how much area is covered, but also how deep it goes. Again, satellite data can tackle this, but only through more sophisticated measurements. Probably the most reliable method available today is provided by NASA's Gravity Recovery and Climate Experiment (GRACE) satellite. Amazingly, this satellite can detect changes in Earth's gravity related to surface-mass variations. There can be all sorts of complications in such a measurement. For example, the Earth's crust can actually rebound upward when heavy ice is removed from the surface (as a result of melting) like a trampoline after a bowling ball is lifted off it. This crust movement must be incorporated into the analysis of the data. That said, scientists have many approaches to measuring land-ice mass, and they all agree within the certainty of the measurements techniques. The current consensus is that the Antarctic ice sheets are losing ice at a rate of about seventy billion (that's with a "b") metric tons per year.* That's enough water added to the ocean each year to fill about twenty-eight million Olympic swimming pools. Even Michael Phelps hasn't covered that many pool lengths. In terms of sea-level rise, we're talking about roughly two millimeters per decade coming from this particular contribution. This ice loss has been accelerating recently, so the rate will surely climb further in the future. And there are more, bigger contributions to sea-level rise coming from other places, as we'll explain later in the chapter.

Brad's back with a wry smile. "So you mean to tell me that ice is melting . . . in Antarctica? Isn't it way too cold there for ice to melt?" Indeed, Brad! Even in summer, the air temperature is typically always below the freezing point of water. But it isn't the air temperature that's melting the ice; it's that warm ocean water we mentioned earlier. The warm water gets in at the edges of the continent and melts the ice from below. This phenomenon is still not fully understood by scientists, but the general view is that winds strengthened by the ozone hole are helping to push the water farther onto the continental shelf, causing the land ice to slide into the ocean. So while Antarctic sea ice is indeed growing—for now—the Antarctic land ice is melting away at an ever-increasing pace.

Let's turn our attention to the opposite end of the planet. The loss of Arctic sea ice (flavor 1) has received a lot of press in recent years (see Chapter 3 for the advantages and disadvantages of this phenomenon), and as we mentioned, the sea ice in Antarctica is growing. Brad seems to think that the gains in Antarctic sea ice make up for the losses in Arctic sea ice. What do the numbers show?

There are a few ways to look at this, but regardless of how you

---

*Interestingly, the East Antarctic Ice Sheet has actually gained some mass (seemingly due to increased snowfall in that region), but the West Antarctic Ice Sheet and Antarctic Peninsula ice sheet losses far outweigh those gains.

measure things, the assertion that the net effect is zero is way, way wrong. First we'll look at the sea-ice extent, which is the area covered by sea ice, in the summer (when the sea ice is at its lowest extent). The Arctic summer is the Antarctic winter and vice versa, so you have to compare different months of the year here. (Believe it or not, skeptics have made bold claims regarding the sea-ice comparison without remembering this simple fact.) Drawing from satellite data going back to 1979, the summer Antarctic sea-ice extent (measured each February) has grown by about 300,000 square kilometers in that time. How about the Arctic sea ice (measured each September)? It has lost about 1,500,000 square kilometers, or about five times as much as the Antarctic gained! If you think that's an even trade, Brad, I've got a bridge to sell you.

Now, the extent of sea ice is important from the standpoint of how much of the sun's radiation gets scattered back into space as opposed to absorbed by the ocean (which directly impacts the pace of global warming), but from a sea-level-rise standpoint, we have to look a little deeper—literally. If you want to know how much ice is really there, you need to figure out the ice volume, not just the ocean area covered by ice. First-year ice is rather thin—sort of a crust on top of the water; it takes many years of ice accumulation to build up thick layers, which scientists have creatively called "multi-year ice." That thin first-year ice is susceptible to all sorts of short-term effects (and when these brief variations happen to involve more ice growing, skeptics jump all over it and say the planet is obviously not warming), so—as always—you can only really tell what's going on with climate change by looking at long-term data—and at the total amount of ice. Measurements taken on the Earth and from satellites (with radar and laser altimetry) agree strongly that, in addition to the loss of sea-ice extent, the multi-year ice volume is dropping precipitously. In other words, the ice is shrinking *and* thinning.

Incidentally, some skeptics who acknowledge the loss of Arctic sea ice attribute it to natural variations and point to melts from the early twentieth century. In reality, the Arctic sea-ice extent is now significantly lower than it has been at any point, not just in the last century but in at least the last 1,450 years (measured by a combination of ice-core, tree-ring, and lake-sediment data). That'd be since the time of the legendary warriors King Arthur and Beowulf—quite a lot of natural variation to see over such a period. There is no doubt among the experts that our activity is driving this change.

So what about land ice (flavor 2) in the Arctic? There is indeed a massive area of land ice in this region, too: Greenland. (Does anyone else find it strange that Greenland is icy and Iceland is green?) Brad is telling you that Greenland is gaining ice, but as is so often the case with these skeptic arguments, he's cherry-picking the data. You see, global warming causes the land ice to thin and break away from the perimeter, much like it does in Antarctica, but it also increases the snowfall in Greenland's interior. If you look just at ice mass in the interior regions (as Brad and friends seem to do), you will sometimes see areas where the mass has increased. When you look at the Greenland ice sheets as a whole, however, there is unquestionably a significant loss in ice mass.

What is truly alarming about this ice loss is that it is accelerating at a pace that exceeds the expectations of models. The current period of loss started in the 1970s, coinciding with the time at which global temperatures began to climb again (see Chapter 7). GRACE data from the past decade clearly show the acceleration of ice loss, with the rate of loss increasing by about thirty billion tons/year each year. Overall, Greenland has lost about two trillion (with a "t") tons of ice in the last ten years—that's far more than from Antarctic ice sheets in the same period. And it's getting faster all the time!

When you put together the ice losses from Greenland and Antarctic ice sheets, the contribution to sea-level rise is expected to be something like 0.8 to 3 meters by the year 2100. Barring expansive and expensive engineering efforts to protect cities, this sort of rise would result in severe flooding at high tide in low-lying cities like Miami, Florida (and Orlando for that matter . . . hope Mickey and Minnie can swim), Singapore, and Tokyo. If you push your time horizon out a little farther, the sea-level rise becomes far more serious—there's an awful lot of water locked up in those ice sheets, and if the planet warms enough, much of it will eventually return to the ocean from whence it came.

What about the glaciers elsewhere on Earth? Brad is keen to point out that he's seen data for glaciers that are growing, which is in fact accurate. Guess what he's done here . . . again? Yup . . . cherry-picked the data. Before we get to the data, though, let's learn some glaciology (yes, that's really a word!). Glaciers may look like fairly static things, but they are constantly changing. Gravity is always tugging on these massive objects, causing them to slowly sink down toward sea level. They gain mass on their upper regions from snow-fall and from meltwater that refreezes, and they primarily lose mass on their lower regions from melting on the surface, melting from below, and the calving of icebergs (those are the dramatic events you may have seen footage of when giant chunks fall off into the sea). Odds are that, for any given glacier, the gain and loss rates will not be exactly in balance (glaciologists would call such a condition the "balance velocity"), which means that the glacier will either be growing or shrinking, and that it will move.

OK, so what's the status of Earth's glaciers overall? Brad has cor-rectly pointed out that some are growing, but the vast majority of them are shrinking.* Moreover, the reason behind the growth of some glaciers is increased snowfall resulting from—another drum-roll please—global warming.

Like the polar regions, glaciers worldwide provide early signs of a changing climate. This is because glacier melting is very sensitive to air temperature. Glaciologists monitor hundreds of glaciers across the planet using a plethora of techniques, and the numbers show that

---

*As discussed in Chapter 2, the Intergovernmental Panel on Climate Change (IPCC) made an unfortunate error in one of their assessment reports in which they repeated an unsubstantiated claim that Himalayan glaciers would be completely gone by 2035, a fact that skeptics have pounced on as evidence for sloppiness or even a conspiracy.

not only are most shrinking but the number of growing glaciers is also decreasing with time. Today, about 90 percent of glaciers are shrinking. Brad can throw the Waldemarbreen or Djankuat Glaciers in your face as examples of growth, but there are so many counter-examples that you'd have to brush up on your Nordic and Russian to even consider replying with a list of names. Not only are all these glaciers shrinking, they are shrinking at an accelerating rate.

The upshot? Skeptics can cherry-pick a few examples where ice is gained (though even in these cases it is often due to global warming), but the complete picture is unambiguous: Earth is losing ice at a rapid and accelerating pace. This phenomenon will cause the sea level to rise substantially in time, and it is a harbinger of far more trouble ahead.

# 9

# Climate is too complex to model or predict

**NIELS HENRIK DAVID BOHR** was a brilliant Danish physicist (awarded with a Nobel Prize in 1922) born at the tail end of the nineteenth century. His contributions to science were monumental in the areas of atomic structure and quantum mechanics, but it's some of his other contributions that we're interested in. You see, Niels was something of a scientific Yogi Berra: a guy who reached the pinnacle of his profession but who's known by many as much for his wonderful quotes as for his professional accomplishments. Among the most delightful is that "prediction is very difficult, especially about the future."*

There are some things that we can predict with near-absolute certainty. Like that the sun will rise in the east next morning, or that the Chicago Cubs will not win the World Series this season. Lots of things, of course, are less certain. In particular, really complex stuff is far more difficult to predict with high certainty, and so our skeptic Brad is, well, skeptical that climatologists have even a slight chance of predicting what will happen to something as complicated as the Earth's climate . . . as Bohr would say, especially in the future.

A common incarnation of this skeptic argument says, "Even with state-of-the-art computer models, scientists can't even accurately predict the weather two weeks from now, so how can they possibly tell us what's going to happen twenty years from now?" Here we have

---

*It turns out that there's a dispute about who actually said this first, but being big Niels Bohr fans, we're giving him the credit.

one of those classic climate-skeptic mistakes. (Nope, not cherry-picking this time, but we're glad you've picked up on that common skeptic mistake.) This mistake is confusing weather with climate. As we explained back in the Prologue, weather is what you get day to day, which is something that can vary considerably based on all sorts of influences. It's chaotic, in the scientific sense of the word. Chaos theory looks at the behavior of dynamical systems that are highly sensitive to initial conditions, and people have applied it to everything from data-encryption systems to models that try to predict epileptic seizures. It is because of the extreme sensitivity to initial conditions that weather predictions are such a challenge. Even a tiny change in these initial conditions can have a dramatic effect on how things unfold, something known as the "butterfly effect."*

Climate isn't the same as weather. Climate is the average of weather over a long time—years or even decades. Think of it in terms of a slot machine at a casino. Predicting weather would be like trying to predict whether you're going to win on the next pull of the lever, which, despite some folks' belief that they can do this, is not going to be successful very often. Predicting climate, on the other hand, would be like predicting whether you'll win or lose money averaged out over thousands of pulls of the lever. As casino balance sheets will tell you, the odds are pretty darn high that you're gonna lose.

Bringing it back to weather and climate: Consider when the temperature one day to the next differs by, say, five degrees Celsius (nine degrees Fahrenheit). You might not even notice such a change. It is an entirely different story if the *climate* changes by those same five degrees Celsius, because a climate change means that the average global temperature has changed. We're no longer looking at a brief, local fluctuation.

What would it look like if Earth were five degrees colder on average than it is now? We would be in a glacial period. Sea levels would

---

*A marvelous 1952 science fiction short story by Ray Bradbury called "A Sound of Thunder" is based on this effect.

be perhaps 100 meters (over 320 feet) lower than they are today.* Places like Chicago, London, and Manhattan would be buried under ice. Some places that are now deserts, like in the American Southwest or Afghanistan, would be wetter, while other deserts would expand. All in all, our planet would be a very different place. What about five degrees in the other direction (warming)? Ice would likely disappear completely from both poles, sending sea level about 20 meters (more than 65 feet) higher than it is today. Oxygen levels in the oceans would drop dramatically, suffocating many higher forms of life in the seas. Oceanic methane eruptions coupled with ignition sources such as lightning could unleash unimaginable destruction. The tropics and sub-tropics would be essentially uninhabitable. Super-hurricanes would wreak havoc on those who managed to survive. Basically, you've got something approaching the biblical apocalypse.

Clearly, five degrees is a pretty serious thing in terms of climate (and just so we're clear about where things stand: five degrees of warming is within the range of possibility this century if we don't start changing our behavior as a society). Weather and climate are really quite different beasts. The real question, then, is not if scientists can model and predict weather, it's if they can model and predict climate.

Brad tells you that climate models are riddled with fudge factors to fit our current climate, and that this explains why they agree with observations today, but that these ad hoc insertions are unlikely to be appropriate for modeling a future climate with different atmospheric chemistry. The point he's trying to make is that climate models' dire predictions of future global catastrophes (see Chapter 3) are melodramatic. This leads to another question: How do you test the accuracy of a model that is projecting trends many years into the future?

---

*The last time this sort of thing happened, the drop in sea level aided human movement from Asia into North America because the land now under Alaska's Bering Strait surfaced to provide something of a bridge.

It's important to keep in mind that we're talking about climate here, not weather. The object is not to predict what temperature it will be in Delhi on, say, March 5, 2023, or exactly when the next hurricane will hit New Orleans. Instead, we want to predict the trend in the climate over the coming decades. Like modeling weather, this, too, is no easy task. Climate models have to account for interactions between the land surface, ice, oceans, the sun, and the atmosphere, among other factors. What's remarkable is that despite their complexity, climate models do an impressively good job, as we'll explain, and are improving all the time.

There are two basic approaches to testing the accuracy of a climate model. You can wait and see what happens in the future and compare it with the predictions of your model, or, if you don't have the luxury of waiting for years and years, you can do what's called "hindcasting." This process involves picking a starting point somewhere in the past, feeding those (known) climatic conditions into your model, and letting 'er rip. If the model projects trends that match what actually occurred from that starting point, you can have reasonably high confidence that it will also do a good job projecting what will happen years from today. It is standard practice for climate modelers to first test their models with hindcasting before using them to project future climate trends, and good models routinely pass this test. Of course, modelers design their models such that they would be expected to match the observational data (a little like knowing the questions on the pop quiz ahead of time), but nonetheless, if they failed to capture historical changes, you wouldn't have much faith in their predictive capabilities.

In fact, hindcasting like this has shown that, unless you incorporate human $CO_2$ emissions and their associated greenhouse effect, you cannot properly simulate recent climate changes. Other, natural factors are sufficient to explain temperature variations prior to the relatively recent warming period, but models without anthropogenic influence fail to reproduce the warming observed since the middle of the last century. This is among the strongest pieces of evidence that global warming is in large part caused by human activity.

There are some models, however, that have been around long enough for us to check in on how they've been doing with their challenge of predicting the future. One such model was launched back in 1988 by a team of NASA and MIT scientists led by James Hansen. The Hansen model predicted a trend of rising global temperatures that does a very good job of tracking the trend that has actually taken place in the decades since it was published. Hats off to Hansen and his team for working on this issue before most of us had ever heard of global warming. (Al Gore, for one, who was a US Senator at the time and busy running for the Democratic nomination for president, was certainly trumpeting environmental issues like the growing ozone hole over the Antarctic,* but he hadn't yet assumed his role as one of climate change's most visible activists.)

It turns out that Mother Nature provided a convenient opportunity on June 15, 1991, to test the Hansen model as well as other climate models. Mount Pinatubo, a volcano located in the Philippines, experienced a massive eruption that day—one of the largest of the twentieth century. This event launched about twenty million tons of sulfur dioxide into the atmosphere. The resulting sulfate aerosols in the stratosphere had a gigantic influence on global temperatures—about 0.5 degrees Celsius (0.9 degees Fahrenheit) of cooling—that lasted about three years. How did climate models do in predicting the effects of the Pinatubo eruption? Strikingly well. Not only did they predict the temperature drop correctly, but they also predicted feedback related to water vapor and radiative effects that quantitatively matched what actually happened.

Climate modeling may be hard, but these guys are good. Brad may not believe in their models, but as Niels Bohr once replied when he was asked by a visitor to his home in Tisvilde if he really

---

*Refer back to Chapter 8 for an interesting connection between the ozone hole and climate change.

believed a horseshoe above his door would bring him good luck, "Of course not . . . but I am told it works even if you don't believe in it."

Probably in part because the projected consequences of climate disruption are so scary, skeptics seem to find welcome ears for their claims that climate models exaggerate the threats posed by our greenhouse-gas emissions. It sure would be nice if that were true. Let's take a look at some model projections and see how they've panned out.

The IPCC's first assessment report was released in 1990. One of the landmark predictions from that report involved sea-level rise. As we've discussed in Chapters 3 and 8, heating of the oceans and melting of ice sheets and glaciers results in the sea level climbing progressively higher. Pulling from a variety of climate models, in 1990 the IPCC predicted a range of possible rises in sea level, with the most probable value being 1.9 millimeters per year. Today, researchers have a great set of data from tide gauges and satellite observations, and they not only agree with each other extremely well, but they also show a trend of rising sea level from the early 1990s through the present time. What do these data show regarding the pace of that rise? Did the IPCC embellish the potential effect? It turns out that the actual rate of sea-level rise during this period was 3.4 millimeters per year. Interestingly, this value aligns with the worst-case scenario from the IPCC report in 1990. In other words, the IPCC didn't exaggerate the situation at all. Quite the opposite. It appears that the IPCC underestimated the scale of the change.

Ice melting is another major indicator of climate change, and so climate models project the pace of things like melting of the Arctic sea ice. When you compare actual observations of this melting with the model projections reported by the IPCC, you find an alarming result. As with the sea-level-rise projections, the IPCC models produce a range of possible rates of sea-ice melting. The sea ice, however, doesn't read IPCC reports. The actual rate of ice melting has been a full 40 percent faster than the average of the IPCC model predictions. In fact, the actual melting rate is faster than the worst-case scenario coming out of those models. The bad news is that the models have failed to quantitatively capture some impor-

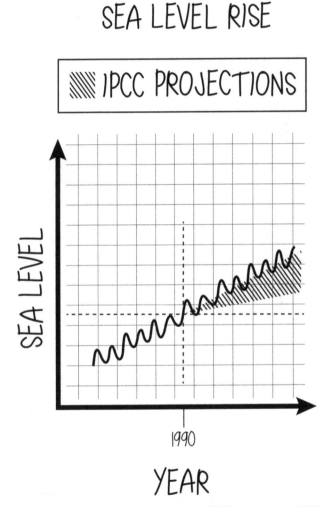

Schematic of sea-level rise showing how the IPCC's projection in 1990 under-estimated the effect

tant effects. The really bad news is that things are even worse than our worst expectations. The IPCC has also underestimated how fast emissions of carbon dioxide would rise, and they've even underestimated the degree of confidence that climate scientists have in the fact that human activity is the primary driver for climate disruption (see Chapter 1). What's the good news? Well, perhaps this sort of thing will make Brad stop saying that the IPCC is exaggerating the problem. The models don't get some things right, but where they get it wrong, it is almost always in the direction of underestimating the scale and pace of the problem.*

We've pointed out the ways in which climate models have gotten things right, and ways in which they've gotten it wrong in precisely the opposite way claimed by the skeptics. There are things that climate models are really good at, but there are also things that give them what could very well be an insurmountable challenge.

In an effort to make climate models more useful, climatologists have worked hard to increase their spatial resolution; that is, to provide predictions for how climates will change on a more regional scale. This is important because, while the planet is warming on a global scale, local changes will vary quite a lot from one region to the next. The problem is that the higher the spatial resolution of your model, the more uncertainty there will be in the projections. Things like the amount and timing of precipitation in a particular area are highly variable. Basically, the models are starting to encroach on something more akin to weather forecasting, which we've already explained is far more difficult than long-term global climate forecasting.

Let's take as an example the Mekong River Basin, a system in

---

*A rare counterexample would be the surface-air-temperature trends in the past decade or so, which have risen more slowly than the vast majority of models had projected (see Chapter 7). Climate scientists are beginning to understand the reasons for this error (volcanic eruptions, unexpectedly low solar output, and so on), but this reminds us that climate models are not necessarily accurate over short timescales.

Southeast Asia that plays a central role in agriculture and trade in the region. A leading climate model projects that the monthly water discharge from this basin could decrease by 16 percent . . . or it could increase by 55 percent. So much for being useful! What's a policy maker or a rice farmer supposed to do with that information? The idea is to feed the results of these regional climate models into so-called impact models that project how the quality of human lives will be changed by variations in the local climate. Not only is the input information uncertain, but it's daunting to try to model the resilience of a particular culture to these environmental changes. This sort of modeling is not really climate science at all. Socioeconomics is perhaps even more complex than climate.

Another way that socioeconomic complexity finds its way into climate modeling is through the one factor that plays the biggest role in changes to our future climate—our own greenhouse-gas and aerosol emissions. As we've already said, the IPCC has badly underestimated the growth in carbon dioxide emissions. This aspect of climate modeling is really economic modeling. Researchers have to be able to predict how we will be using fossil fuels over the course of many decades in the future. How do you model how quickly societies will shift to alternative energy sources? Or even the health of the global economy on that timescale, which has a direct impact on the amount of emissions? If the recent Great Recession taught us anything, it's that economic prediction is a perilous task. Climate models will always get better over time, but that doesn't mean that the uncertainty in their projections will decrease, because human behavior is impossible to predict with certainty.

Another critical area in which climate models are especially challenged is in capturing the effects of tipping points. These are potentially lightning-fast changes such as the much-talked-about melting of the Greenland Ice Sheet, which, if it were to completely and rapidly melt, would raise the seas to a frighteningly high level. The problem is that climate models are designed to be stable. You wouldn't want your simulation running off into oblivion—that is, unless running off into oblivion is what's actually going to happen! These rapid changes are not well understood, and thus they're not well integrated into models. When climate models are tested against past abrupt changes in climate using hindcasting, they generally don't do a good job of reproducing these cataclysmic events. What this means is that climate models will generally get the right answer, it's just a question of the timing. Models can tell us with very high certainty that specific levels of global warming will occur . . . they just can't tell us, with an equally high certainty, *when* those levels will be reached.

All right, so we have a sense now for what climate models can and can't do well. Where do we go from here? What are policy makers—and the rest of us—supposed to do? Brad wants us to wait until scientists have proved that it's our activity that's causing climate change. Check (see Chapters 1, 10, and 11). Then he says, "We need to wait until the modelers can tell us exactly what's going to happen and when." No, Brad. We can't wait for that, because the models will never do that and we'll still be waiting while catastrophic effects begin to destroy our society. There is enough certainty now to act. We've created an awful mess for ourselves as a species, but there are things we can do to prevent the worst consequences from taking place. This will require public support and government action, and the skeptics are doing their best to stymie both of those things.

The gravity of the predicament we've gotten ourselves into as a species is so grim that, at times, we feel guilty about poking fun at the most audacious and absurd of Brad's claims. At times like these, we return to our old friend Niels Bohr for some solace. He once sagely declared, "Some subjects are so serious that one can only joke about them."

**4**

# It's Not Our Fault

There is no link between carbon dioxide and global temperature

It's just a natural cycle

More carbon dioxide won't make a difference

What about . . .

# 10

# There is no link between carbon dioxide and global temperature

**A CORNERSTONE OF** the concept of anthropogenic climate change is that carbon emissions from fossil fuels lead to global warming via the greenhouse effect. It shouldn't be surprising, then, that skeptics try to discount the connection between carbon dioxide and global temperature. Skeptic arguments come at this topic from many different angles, and we'll address each of them in this chapter. As a rule of thumb, in each case skeptics make mistakes of the nature you've seen many times in this book already: They cherry-pick data and confuse timescales. Let's start with a glaring example of the latter.

As we discussed back in the Prologue, scientists have reliable data on Earth's climate history dating back hundreds of thousands of years, thanks to deep Antarctic ice cores. Over this long period, the planet has cycled between rather long Ice Ages and relatively short warmer periods in between called interglacials. A complete cycle takes about one hundred thousand years. When you look at these data, you find a striking correlation between carbon dioxide levels and temperature throughout the historic record. If you look closely, however, you'll see that while these two parameters rise and fall together, there is a significant time lag between them. Specifically, the carbon dioxide changes typically occur hundreds of years after the temperature changes. Because of this time lag, Brad feels confident that carbon dioxide doesn't cause global warming . . . he says it's the other way around! As it turns out, he's half right. $CO_2$ and temperature play off each other in complex ways, each influencing the other like "Dueling Banjos" from the movie *Deliverance*, except without the creepy hillbillies.

To understand the historical connection between carbon dioxide and temperature, we need to look at the driving forces for those ice age/interglacial cycles. It all starts with some astronomy and a Serbian scientist named Milutin Milanković. He was held prisoner in the early part of the twentieth century as a result of the conflict between the

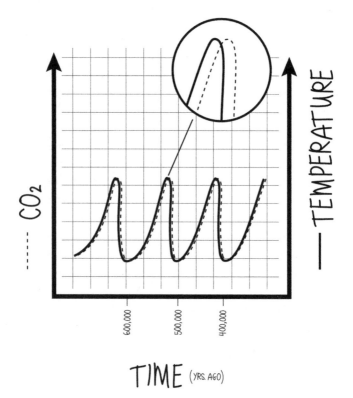

## ICE-CORE RECORDS

Schematic showing the temporal relationship between carbon dioxide and temperature in the ice-core record

Austro-Hungarian Empire and Serbia (the one that led to the First World War). Thanks to some powerful friends, he was permitted to spend his captivity in Budapest and to continue his work. It was in this time of imprisonment, during which he must have had plenty of time to think, that he developed the idea of a connection between orbital motions and the Earth's climate. In his honor, the phenomenon describing this relationship is known today as Milankovitch cycles.

There are three ways in which the Earth's orbit changes over time: (1) The axis around which the planet spins wobbles slowly, or precesses,* a process taking about twenty-six thousand years per cycle; (2) the angular tilt of the Earth's rotation axis, known as its obliquity (awesome word!), varies from about 22 degrees to about 24.5 degrees, a process taking roughly forty-one thousand years per cycle; and (3) the shape of the Earth's orbit around the sun is not a circle. Rather, it's an ellipse, and the eccentricity (non-circleness) of this ellipse varies, a process taking somewhere around one hundred thousand years per cycle.

Other than being a mathematical and scientific triumph, these Milankovitch cycles also tell us something important about our climate. Together, the three cycles help to determine how much sunlight hits the Earth—especially closer to the poles. The direct effect of this change in sunlight is relatively small, but it triggers a series of events that end up having dramatic influence on the climate.

For example, there is a point in the cycle at which orbital changes cause warming at high latitudes, leading in turn to melting of ice sheets. All that freshwater ends up in the oceans, which disrupts the ocean circulation patterns. As a result of these changes, the Southern Ocean warms up, which decreases the amount of carbon dioxide that the ocean can dissolve, thereby causing the emission of excess $CO_2$ into the atmosphere.

---

*This precession means that the North Star is not always the same star. Today it is Polaris in the constellation Ursa Minor, but other times the axis points to stars like Thuban in the constellation Draco.

# MILANKOVITCH CYCLES

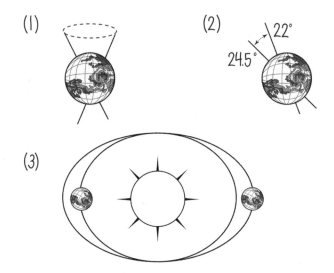

**Three sources of Milankovitch cycles: (1) precession, (2) obliquity, and (3) eccentricity**

So there you go, Brad . . . the temperature increase does indeed drive a carbon dioxide rise. But that's not the end of the story. That carbon dioxide released from the ocean greatly amplifies the (initially mild) warming via the greenhouse effect. The bulk of the actual warming that shifts the planet from a glacial period to an interglacial ends up occurring after the $CO_2$ release. Warming begets $CO_2$, which begets additional warming. It's a positive feedback cycle. That explains the time lag in the ancient climate data. But what does the correlation between carbon dioxide and temperature look like in more recent times?

It's important to distinguish between natural and anthropogenic influences on the global temperature, collectively referred to as "forcings." There are many forcings that affect our climate—we'll spend more time on some of these in Chapter 13—and not all of them push temperatures higher. Some forcings, both natural and man-made, lead to a cooling effect. Examples of cooling forcings include things like aerosols in the atmosphere (both natural and man-made), the presence of stratospheric ozone, and surface albedo (how reflective the planet's surface is to incoming sunlight). Warming forcings include tropospheric ozone, stratospheric water vapor, airplane contrails, and of course solar irradiance and greenhouse gases. Because of this menagerie of forcings pushing this way and that, short-term variations in the global average temperature can be substantial and confusing, but the long-term trend is really what we're interested in when discussing climate change.

Brad points out that carbon dioxide emissions rose only slightly during the beginning of the twentieth century and yet there was warming during that time; moreover, he says, emissions rose substantially after the Second World War, yet the temperatures actually cooled slightly in the middle of the century. These periods of poor correlation between carbon dioxide emissions and temperature seem to contradict the fundamentals of the greenhouse effect—unless, that is, you consider all the other forcings taking place during those times.

A common misconception (though not among climate skeptics, of course) is that anthropogenic climate change has been the dominant climate influence since the Industrial Revolution. Actually, greenhouse warming from human activity, though a contributor since that time, has only become the dominant contributor since the 1970s or so. In earlier years, other factors were stronger than the comparatively small forcing from carbon emissions. These other factors can explain the periods of poor correlation between carbon dioxide and temperature in the last century.

Even when you look at the period of warming that started in the 1970s (and that correlates with increases in carbon dioxide emissions), you find stretches where cooling takes place over a few years (like 1977–1985). These short-term changes do not represent real climate change. To be considered a real climate trend, a change has to occur over a period of at least thirty years (see the Prologue). The main driver of these shorter-term variations in climate is ocean cycles like El Niño/La Niña that move heat between the water and the atmosphere. Solar cycles, in which the sun itself releases greater or smaller amounts of radiant energy, can also impact the climate. If you happen to get a period of low solar intensity at the same time as a La Niña (this happened in 2008), these two effects can gang up on greenhouse warming and beat it down for a little while. In the end, however, the long-term trend overcomes such variability.

Unlike the aforementioned brief climate deviations, the mid-century cooling period was actually long enough to be considered a change in climate, but one that has an interesting explanation fully consistent with anthropogenic climate change . . . check back in Chapter 7 for the story. The warming period at the beginning of the twentieth century was also long enough to be a climate change, though the skeptics tend not to focus as much on these periods when the warming is even more than anticipated from the greenhouse effect. At that time, roughly 1910–1940, the sun was on an upswing in its energy output, there were statistically few volcanic eruptions (which have a cooling effect by scattering away sunlight back into space before it reaches the Earth's surface), and the ocean cycles (the Atlantic Multidecadal Oscillation in this case) were contributing to warming as well. Taking these forcings together with the greenhouse warming from carbon dioxide emissions at the time (which were much lower than today's), you get a good rationale for the observed warming. In the period since the 1970s, with much higher carbon dioxide emissions, variations in solar activity and the like can no longer completely mask the greenhouse warming. Moreover, the average annual warming observed in the past forty years has been significantly faster than that during the beginning of the twentieth century.

How much $CO_2$ have we added to the atmosphere? Before the Industrial Revolution, the concentration of carbon dioxide in the air was about 280 parts per million (ppm). As of today, the concentration is about 400 ppm and climbing, already more than 40 percent higher than it was. This has lots of folks concerned.

Brad, however, is not so worried. He claims that back at the end of the Ordovician era (that's about 444 million years ago . . . dinosaurs didn't show up for another few hundred millions years), the $CO_2$ level was over 5,000 ppm! Was the planet a scalding-hot wasteland? Nope. Brad correctly points out that this period was marked by large-scale glaciation. Seriously. What gives?

First off, the carbon dioxide estimate can be misleading. In the deep past, like in this period, we only have data points spaced about ten million years apart from one another.[*] As you can imagine, climate and $CO_2$ levels can change quite a lot between those data points, particularly in the context of a period of glaciation that lasted "only" a million years. Most likely, the carbon dioxide level was substantially lower than 5,000 ppm during the time of glaciation, but nonetheless, it was certainly much higher than today's $CO_2$ levels. So the question remains: How can you have an Ice Age with greenhouse-gas levels so much higher than we have today, if we say today that even a little more $CO_2$ could cause catastrophic climate change? It all comes back to those other forcings we mentioned earlier.

Carbon dioxide levels were higher back then, but the sun's output was lower, and the sun's irradiance is most definitely a major climate forcing. For a given intensity of sunlight (and holding other forcings constant), there is a threshold level of $CO_2$ above which ice cannot survive for long because of greenhouse warming. Less sunlight means a higher $CO_2$ threshold for ice melting, and vice versa. Today, the $CO_2$ threshold number is about 500 ppm (hence the concern that we've already surpassed 400 ppm). The threshold 444 million years ago, when the sun wasn't as bright? A whopping

---

[*]These come from GEOCARB, a geochemical model of ancient carbon dioxide levels developed by a team at Yale led by Robert Berner. Notably, Berner has specifically stated that his model should not be used to gauge what the carbon dioxide levels were in the Late Ordovician era because of the large timesteps of the model.

3,000 ppm! That's an impressively high threshold for ice to still be happy and frozen, but it is below the 5,000 ppm data point. So could the $CO_2$ level have dropped from more than 5,000 ppm to less than 3,000 ppm during this period of glaciation? Scientists are able to piece together information about the carbon dioxide level in between those ten-million-years-apart points by looking at other proxies for $CO_2$,* and these studies suggest that the answer is yes.

So all of the data, from hundreds of thousands of years ago to the present day, are consistent with the consensus view of the role of $CO_2$ in climate change. Up until extremely recently, the interplay of $CO_2$ and climate was a purely natural phenomenon, with climate changes taking place over thousands or millions of years. It is only in the past century that we have begun to tinker with that natural system by rapidly putting carbon into the atmosphere, carbon that had been trapped in fossil fuels for millions of years. We'll spend the rest of this chapter exploring these emissions caused by human activity and some skeptic myths related to them.

First things first: Is the level of carbon dioxide really climbing? Probably the most foundational data underlying anthropogenic climate change is known as the Keeling Curve. It's named after Charles David Keeling, who worked at the Scripps Institution of Oceanography at the University of California, San Diego. He was the first scientist to perform regular measurements of atmospheric carbon dioxide concentration, starting in 1958, and his now-famous paper published in the meteorology journal *Tellus* in 1960 was the first to show empirical evidence that $CO_2$ levels were rising in accordance with global fossil-fuel use. The longest uninterrupted set of measurements, which shows a continuous increase in $CO_2$

---

*One example of a $CO_2$ proxy is the study of rock weathering, a process that removes carbon dioxide from the atmosphere. Rock weathering produces a strontium isotope that gets washed into the oceans and trapped in sediments, where it can be measured today.

concentration from 1958 until today, is from Mauna Loa in Hawaii.[*] The numbers started around 315 ppm at the beginning and are now in the neighborhood of 400; the Keeling Curve is the graph of these data, which makes an appearance in the sharp upswing of the graph of Mike's Nature trick on page 19.

Brad scoffs, "So you mean to tell me that your Keeling Curve data are taken from a volcano? That seems a poor choice for sampling atmospheric gases."

While it may be true that active volcanoes can spew carbon dioxide into the atmosphere, the fact is that carbon dioxide mixes into the atmosphere very well, so the measurements are essentially the same worldwide. In fact, there are now dozens of apparatus around the globe (and even on satellites) measuring $CO_2$ concentration, and if you look at the average of all these sites, they align with the Mauna Loa measurements. In other words, these data are highly reliable. As striking as the rise in the Keeling Curve is, it is all the more arresting when you attach these data to the historic record from ice cores. When you view it all together, you have a clear perspective on how rapidly we are changing the atmospheric chemistry, compared to natural cycles. This combined data set, when graphed, makes the shape of a hockey stick because the $CO_2$ concentration goes from virtually flat for thousands of years to a sudden, unabated rise initiated right around the start of the Industrial Revolution in 1760.

Any skeptic who claims that carbon dioxide levels are not rising is simply ignoring the basic facts. There is no question that this is happening. The question then becomes: If the carbon dioxide concentration is rising, is it due to natural or human causes? One thing is clear: We are putting lots of carbon dioxide into the atmosphere.

---

[*]The longest set of *direct* measurements, that is. As we've mentioned previously, ice cores and other historic records provide reasonably good measurements of carbon dioxide levels going back into deep history. Keeling actually started measurements at the South Pole, too, but funding cuts in the 1960s forced him to abandon that location.

That's a simple conclusion based on the chemistry of burning fossil fuels. If you burn a hydrocarbon like natural gas (which is methane, $CH_4$) in air, you get two major products: water and carbon dioxide.* We actually know with pretty high accuracy the amount of $CO_2$ emissions worldwide, which was about 33 billion (with a "b") tons in 2008 and has risen substantially since then. All that $CO_2$ has to go somewhere, and the $CO_2$ concentration in the atmosphere has increased in virtual lockstep with the increase in $CO_2$ emissions. Not all of the carbon dioxide stays in the atmosphere, though. A large fraction of it is absorbed by carbon sinks like the oceans and plants, causing havoc beyond climate change (see Chapter 3). The part that remains in the atmosphere is called the "airborne fraction," and it sits consistently right around 43 percent of the total emissions. How do we know this? Time for a little atmospheric CSI (CBS's next spin-off).

Carbon comes in different isotopes. You're probably familiar with carbon-14, the long-lived radioactive isotope used for dating ancient human artifacts. There are actually fifteen known isotopes of carbon, but the two stable ones are carbon-12 ($^{12}C$) and carbon-13 ($^{13}C$), which occur naturally in a ratio of about 99:1. Plants find it easier to use the lighter isotope (carbon-12) when they convert sunlight and carbon dioxide into food via photosynthesis, so plants have a richer ratio of $^{12}C/^{13}C$ than the atmosphere does.

What is coal made of? Plants. Same goes for a good portion of natural gas. The carbon in these fossil fuels, then, will have a lower ratio of carbon-13 than the atmosphere. It's that same carbon that ends up in the carbon dioxide when you burn the fossil fuel. This means that, if the rising carbon dioxide levels in the atmosphere are coming from the burning of fossil fuels, the ratio of carbon-13 to carbon-12 in the atmosphere should be falling. This is exactly

---

*There are also plenty of products produced in smaller quantities, like nitrogen oxide, carbon monoxide, ozone, and so on, many of which cause problems of their own.

what is observed! The $^{12}C/^{13}C$ ratio tracks $CO_2$ emissions from the early 1980s (when people first started measuring them) to today quite nicely. The atmospheric CSI team has found the smoking gun. Fossil fuels plead guilty, your honor.

We've addressed the question about whether the $CO_2$ rise is due to natural or human causes. So now the question becomes: Does the airborne fraction of human carbon dioxide emissions have any effect on the climate? Skeptics argue that the amount of $CO_2$ from human activity is so small compared to the amount in the overall system that it can't possibly make any impact.

While natural carbon dioxide levels have changed dramatically over the course of the Earth's history, they have actually remained pretty constant over the past several thousand years. This doesn't mean that the carbon dioxide in the atmosphere is just sitting

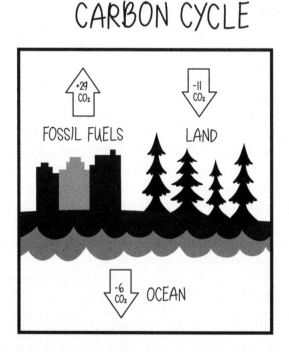

around, however—it's actually pretty busy stuff. Lots of natural processes are releasing $CO_2$, and others are absorbing it, all the time.
The oceans both absorb and release carbon, vegetation is eaten
by animals and microbes that subsequently release carbon, plant
growth consumes carbon, and plant respiration releases carbon.
This movement of carbon dioxide is known as the global carbon
cycle, and the amounts of $CO_2$ involved are enormous.

You can see from the picture on the previous page that the oceans
and plants are absorbing more than they are releasing. Where is that
extra carbon coming from? From us, of course, or our power plants
and cars, and so on, to be precise. That's the part of our emissions
that *isn't* the airborne fraction—it's not in the air if it's absorbed
by oceans and vegetation, after all. A little arithmetic shows that
some of our emissions are left over, and that part—the airborne
fraction—is indeed enough to have an effect. Natural carbon dioxide levels generally take about ten thousand years to change by 100
ppm. We've seen that much increase in the past century or so. Our
activity is disrupting the natural carbon cycle.

A related skeptic myth is that carbon dioxide is just a trace gas in
the atmosphere, which is made up mostly of nitrogen and oxygen.
It's true enough that we're talking about parts per million here, but
that doesn't mean that carbon dioxide is an irrelevant constituent of
the air. If Brad thinks a few hundred ppm isn't anything to concern
ourselves with, maybe he'd be willing to have a few ppm of arsenic
in his morning coffee. Or good luck arguing with the police when
they arrest you for "only" having 800 ppm (0.08 percent—the legal
limit in the United States) of alcohol in your blood while driving.
The nitrogen and oxygen that make up most of our atmosphere play
essentially no role in regulating the temperature on the planet, so it
is irrelevant that the *relative* amount of $CO_2$ is so much lower than
that of those gases. It is the *absolute* amount of carbon dioxide that
matters, and that amount is rising, thanks to us.

Brad's got still one more argument along this line. He says that
carbon dioxide only sticks around for about five years in the atmo-

sphere (true!), so long-term warming from the $CO_2$ that we put up there is nothing to worry about. This reasoning is predicated on a simple misunderstanding of the carbon cycle. The specific $CO_2$ molecules that come out of the smokestacks and exhaust pipes stay in the atmosphere for only a few years on average, but they don't vanish after that time; rather, they end up in other parts of the carbon cycle (most likely absorbed in the ocean or by a plant). Other $CO_2$ molecules are swapping places with those $CO_2$ molecules all the time, but the total amount in the cycle is increasing because of our emissions. The question isn't how long an individual carbon dioxide molecule stays in the atmosphere; it's how long the extra population of molecules sticks around. The answer to that is at least hundreds of years, and that's the period when our climate will feel the warming effect.

So how big is that warming effect from the carbon dioxide? We mentioned climate forcings earlier in the chapter. Let's take a look at some of the numbers for those now, for perspective. The strength of a forcing is measured in units of watts per square meter ($W/m^2$). Each of the forcings has some uncertainty associated with it, but thanks to years of meticulous research, most of these uncertainties are now quite small (with the exception of aerosols, which remain an extremely challenging subject in climate science). Aerosols contribute the strongest cooling effect, with the current estimate being $-1.2 W/m^2$. Another cooling forcing is reflection of radiation off the Earth's surface and back into space, to the tune of about $-0.2 W/m^2$.*
Ozone can cool or warm, depending on where it sits in the atmosphere, but overall its influence is a positive (heating) forcing of about $0.3 W/m^2$. By far the strongest positive forcing comes from greenhouse gases, which include carbon dioxide, but which also include things like methane, nitrogen dioxide, and halocarbons.†

---

*Making surfaces darker, such as by polluting the air with black carbon that settles on otherwise white snow, has a warming effect. Same goes for melting of sea ice.

†These gases are far stronger greenhouse gases than $CO_2$—in some cases thousands of times stronger, but there is much less of them in the atmosphere.

How big is the $CO_2$ contribution among these various gases? Well, all the others together have a forcing of about 1 W/m². Carbon dioxide alone has a forcing of roughly 1.7 W/m². Not only is that the strongest greenhouse-gas effect, but it is also stronger than all the other positive forcings combined!

The way carbon dioxide drives climate disruption (and ocean acidification) has not gone unnoticed, of course. This correlation, for example, has led the US government to recently classify $CO_2$ as a pollutant, much to the chagrin of climate skeptics and the fossil-fuel industry.

Brad sneers, "If you think carbon dioxide is a pollutant, you'd better stop breathing, because you're exhaling air with several thousand ppm of $CO_2$ in it with every breath!"

We, along with all other aerobic creatures, do indeed take in oxygen and give off carbon dioxide. But we are just part of the natural carbon cycle. The carbon in that carbon dioxide we breathe out came from the food we ate, which itself took it out of the air—most likely in the very recent past. We aren't adding any carbon to the carbon cycle—we are just cycling it.

A pollutant, on the other hand, is a substance or form of energy that is added to an environment at a rate faster than it can be dispersed or removed in a harmless form. Carbon dioxide coming from human activities like burning fossil fuels and manufacturing cement clearly fits this definition. We are adding $CO_2$ to the environment at a pace faster than the system can get rid of it, and if you think all that extra $CO_2$ is harmless, we recommend that you reread Chapter 3.

# 11

# It's just a natural cycle

**IN CHAPTER 7,** we covered the skeptic myth claiming that the Earth is not actually warming up. There's another camp of skeptics who acknowledge that the Earth is warming up, but who claim that the changes we are seeing in the climate are purely the result of natural cycles or variations, as opposed to being driven by human activity. This belief is held by many who don't necessarily consider themselves skeptics of climate change. We all learned about climatic variations in school, like the various Ice Ages that the planet has experienced. It seems reasonable to ask if recent changes are simply signs of such natural variation. We know that, in fact, natural forcings do play a role in determining our climate, such as the orbital changes we discussed in Chapter 10 and the El Niño/La Niña Southern Oscillation. But what are the nature and scale of that role relative to anthropogenic contributions? We'll look at these issues in this chapter, starting with the most common natural forcings that skeptics point to as explanations for observed global warming, and we'll end with a summary of how as a society our fingerprints are undeniably all over climate change.

The sun is certainly a natural thing, and skeptics frequently point to it as an explanation for climate changes. Brad says, "Almost all the energy on Earth comes from the sun, right? Well, the output from the sun changes over time, so there's your explanation for why we're warming up."

Over the past thousand years or so, global temperatures have in fact tracked solar activity pretty well. The sun may look steadfast and unwavering up there in the sky (our lawyers have advised us to say that we are not suggesting that you actually look at the sun), but it's really a violent, capricious beast.

A German scientist, Heinrich Schwabe, who worked in the early part of the nineteenth century, was hunting for a planet whose orbit was supposed to be even closer to the sun than Mercury's. (This planet was to be called Vulcan—live long and prosper, Herr Schwabe.) Since Vulcan's location would be so close to the (very bright) sun, he figured that his best chance to spy it would be when it passed in front of the sun, appearing as a tiny dark spot. This search led Schwabe to start carefully studying sunspots, hoping to identify Vulcan among them. As with many great discoveries in science, his initial experiment failed miserably, since there is not actually a planet there to find, but he did stumble upon something else. It turns out that the sunspot activity seemed to have a cycle. The period of this cycle, which also correlates with solar-flare activity, is about eleven years, although it isn't perfectly regular.

It wasn't until the beginning of the twentieth century that some other scientists, led by the Chicagoan George Ellery Hale, figured out the origin of the solar cycle. Hale showed that sunspots come in pairs and are actually magnetic features.* He figured out that the solar cycle is a magnetic cycle with a period of twenty-two years on average. The reason that the period is twice as long as the cycle for solar-flare activity is that the magnetic field flops from north to south, or from south to north, every eleven years on average (therefore taking twenty-two years to complete a cycle), but sunspots don't seem to care much in which direction the field is pointing. This eleven-year cycle is actually just one part of solar variation, which also has more random fluctuations—and these are the variations of greatest interest if you're looking for long-term effects on our climate. A common technique when looking at the output

---

*This was the first detection of magnetic fields outside the Earth, which—believe us—was a big deal at the time.

from the sun is to take an eleven-year average of solar activity. This removes the primary magnetic cycle from the picture so you can see the underlying trends.

It is these trends that have loosely matched global temperatures dating back centuries—that is, before the Industrial Revolution and the concomitant growth in carbon dioxide emissions started playing in the game. In the early 1900s, the sun was in a period of increasing activity, which played a role, along with greenhouse warming from carbon dioxide emissions and low volcanic activity, in the warming during that time (roughly 40 percent was from the sun's activity change). The period we're most interested in, however, is from about 1975 until today, because this is the time frame during which temperatures have really started to climb.

So what has that fickle sun been doing since the late 1970s? When you average out those eleven-year cycles, the sun has been getting slightly *less* active! That's right, the relative influence from the sun has been a minor *cooling* effect on our climate in recent decades.* There are, in fact, so many studies showing that the sun is not the explanation for recent warming that it led climate scientist Ray Pierrehumbert (another Chicagoan!) to say about the "it's the sun" argument, "That's a coffin with so many nails in it already that the hard part is finding a place to hammer in a new one."

"OK," Brad says, "so it's not the sun. But there are plenty of other natural cycles that can explain global warming. Like Dansgaard-Oeschger events."

Like what? No, that isn't Brad making stuff up. Willi Dansgaard and Hans Oeschger were climate scientists who documented a

---

*The solar activity that we are talking about here is the total amount of irradiance coming from the sun. It is likely that these changes in irradiance are not uniform across the solar spectrum; that is, maybe certain parts of the spectrum (say, the ultraviolet part) are changing in different ways than other parts (say, the infrared part) over time. Such variations could lead to complex influences on our climate, because different parts of the incoming sunlight are absorbed by different parts of our atmosphere. That said, it's plenty clear that the sun's output is not the reason for recent global warming.

series of abrupt changes in Earth's climate by looking at Greenland ice cores.[*] These "D-O" events are quasi-regular, appearing about once every fifteen hundred years, and they involve rapid warming episodes in the Northern Hemisphere, typically taking just a few decades, followed by gradual cooling over a longer period. Scientists are still debating the origin of D-O events—they might be caused by solar variations, orbital variations, and/or ocean oscillations. Regardless of their origin, these sound like something that might explain what we're experiencing today, right? Not so fast.

First off, the last D-O warm period took place during medieval times, about one thousand years ago (not fifteen hundred). So we're not due for another D-O warming for another five centuries, give or take. Second, note that we said D-O events involve warming in the Northern Hemisphere. The other side of the planet actually cools down during these events. In other words, the heat on the planet is redistributed, but it doesn't actually increase much overall. That is in contrast to global warming, which by definition involves an increase in the overall average temperature of the planet. Seems awfully unlikely that the current observed warming has anything to do with D-O events. And, pointedly, both Dansgaard and Oeschger were leading voices in connecting carbon dioxide emissions and global warming.

So are there some other natural variations that might explain climate change?

Here's the deal. Scientists can explain the rising global temperatures only by combining the known natural forcings *and* anthropogenic greenhouse emissions, with the latter becoming increasingly dominant with time. Is it possible that some other, as-yet-unknown natural forcing exists out there that could provide an alternate explanation? Yes. That chance will always exist. The important question to ask and answer is: How likely is that possibility?

---

*Chester Langway Jr., from the US Army Cold Regions Research and Engineering Laboratory, also participated in this work, but as often happens with less-famous participants in science, his name wasn't enshrined in the climate-science nomenclature.

A natural forcing strong enough to discount anthropogenic climate change would have to explain not only the observed temperature changes (both the amount of warming and the rapid pace of the change), but also why the greenhouse effect *isn't* causing the warming. There are many pieces of evidence that point to human carbon dioxide emissions being the explanation, and any alternative explanation would have to fit all of these different fingerprints of human activity. The likelihood of some unknown phenomenon fitting so many different phenomena, while possible, is minuscule.

There is a guiding principle in science known as Occam's razor, named after William of Ockham,* an English friar who developed the underlying philosophical concept in the early fourteenth century. The principle of Occam's razor is that the simplest explanation is usually the correct one. Karl Popper, one of the most famous philosophers of science in the twentieth century, put his own gloss on this when he suggested that we prefer simpler theories to more complex ones "because their empirical content is greater; and because they are better testable."[†] Let's put anthropogenic climate change to the test and see how it holds up.

Our starting point is that we're putting lots of carbon dioxide into the atmosphere. Burning of fossil fuels represents the biggest share of these emissions, but there are other sources, too, like cement and steel manufacturing. You can prove to yourself that we're adding $CO_2$ to the atmosphere with some simple math—think of it like balancing your checkbook. The change in your account balance (amount of carbon dioxide) will be the difference between your deposits (emissions) and withdrawals (absorption). Handily, accountants (climate scientists) keep track of these things for us. There are natural and human emissions and absorptions, though the human contribution to absorption of carbon dioxide is

---

*Why the different spelling? No idea.

†Karl Popper, *The Logic of Scientific Discovery*, (London: Routledge 1992).

effectively zero. The account balance (carbon dioxide in the atmosphere) is rising at roughly fifteen billion tons per year. Our deposits (human emissions) are roughly thirty billion tons per year. We know both these things with high certainty. Putting this together, we have a balance of fifteen that must equal thirty plus natural emissions minus natural absorption. Rearranging a bit, we're left with natural absorption equaling natural emissions plus fifteen. In other words, Mother Nature is absorbing much more than she is emitting—she can't be the source of the extra carbon dioxide.

$CO_2$ in the atmosphere = total emissions − total absorption

$CO_2$ in the atmosphere = human emissions + natural emissions − natural absorption

15 = 30 + natural emissions − natural absorption

natural absorption = natural emissions + 15

As we explained in Chapters 3 and 10, the observed rise in atmospheric carbon dioxide levels has been in virtual lockstep with the rise in human carbon dioxide emissions.* And if that weren't evidence enough that the rise in $CO_2$ is because of us, isotopic analysis of the carbon in the atmosphere clearly shows that the additional carbon dioxide is coming from plant-based sources—like coal.

We didn't mention it before, but isotopic analysis of coral reefs provides the same result. Corals draw on dissolved carbon in the water to grow their skeletons, building layers upon layers something like the rings of a tree. Remember that the oceans are a major carbon sink, sucking up lots of the carbon dioxide from the atmosphere, so some of the carbon from our burning of fossil fuels ends up in these coral skeletons. By measuring the ratio of carbon

---

*Skeptics sometimes argue that $CO_2$ rise is an effect of rising temperature rather than a cause of it. One way we know this isn't the case is that temperature rise is jumpy, moving up and down a lot (though clearly up in the long term), yet carbon dioxide levels have been climbing quite smoothly—in keeping with the smooth rise in human $CO_2$ emissions. The story is a bit more complex in the deep (pre-human-influence) past (see Chapter 10).

isotopes in the coral, scientists can demonstrate that our societal carbon footprint is a marker of our impact on the amount of carbon dioxide in the carbon cycle.

Still not convinced that we're changing the chemistry of the atmosphere, Brad? Well, there's more.

When you burn fossil fuels, the combustion reaction consumes oxygen from the air—that's where the $O_2$ part of $CO_2$ comes from. It turns out that not only are carbon dioxide levels in the atmosphere rising, but oxygen levels are also decreasing, and the rate of this decrease matches quite nicely with the burning of fossil fuels. The case is closed on this one. Our activity is unquestionably throwing substantial amounts of carbon dioxide into the atmosphere.

Point number two is that carbon dioxide is a greenhouse gas, which means that this extra carbon dioxide is causing not just any old type of warming—it's causing greenhouse warming. There are some signatures of greenhouse warming that enable us to tell the difference between it and, say, extra warming coming from changes in solar activity. Two of these signatures directly show an increased greenhouse effect (extra infrared thermal radiation being sent back down to Earth), but you'll have to wait until Chapter 12 for the details on these, since they are intimately tied to another common skeptic myth.

If global warming were coming from increased solar activity, we would expect the warming trend to be strongest during the daytime because, obviously, that's when the sun is shining. But if global warming were coming from increased greenhouse warming, we would expect nights to warm faster than days. The greenhouse effect operates all the time—not just during daylight hours. The data over recent decades show that the trend is one of warming for both days and nights (and fewer cold days and nights), but with a notably larger increase in the number of warm nights versus warm days (and a larger decrease in the number of cold nights versus cold days). Greenhouse warming it is. Changes in solar activity simply don't fit the observations.

The explanation for why the greenhouse effect would make nights warm faster than days, rather than at the same pace, is complex, but

# INDICATORS OF HUMAN FINGERPRINTS ON CLIMATE CHANGE

CARBON EMISSIONS

MORE FOSSIL CARBON IN AIR

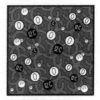

MORE FOSSIL CARBON IN CORAL

LESS OXYGEN IN AIR

NIGHTS WARMING FASTER THAN DAYS

LESS HEAT ESCAPING TO SPACE

MORE HEAT RETURNING TO EARTH

COOLING STRATOSPHERE

RISING TROPOSPHERE

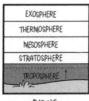

SHRINKING THERMOSPHERE

changes in cloud cover and other cloud properties play a central role. Let's take a moment to talk about clouds, since they are a powerful natural player in the climate game and one that global warming is changing.

The amount of water vapor in the atmosphere exists in direct relation to the temperature. If you increase the temperature, more water evaporates and becomes vapor. Since water vapor is a greenhouse gas, this additional water vapor causes the temperature to go up even further—a positive feedback. But water vapor is "lost" when clouds are formed. This is because water vapor is converted to liquid cloud drops or ice crystals. Some clouds result in precipitation, and the water is returned to Earth, while other clouds form that never result in precipitation. (All clouds eventually evaporate.)

The form of water in clouds (liquid or ice or a combination), the type of cloud,* the cloud location, and the time of day or night when clouds are present are all important factors in determining if a cloud will warm or cool the planet. Considering that clouds cover more than half the globe during any given month, clouds are really important to characterize and understand.

While much about clouds remains poorly understood, it is known that they play a powerful role in regulating our planet's temperature. Even a 1 percent change in cloud cover can have dramatic effects. Clouds have a dual role in balancing Earth's heat budget. On the one hand, clouds spread out like global parasols, blocking the sun's radiant energy and reflecting much of it back to space, which has a cooling effect. On the other hand, like $CO_2$, clouds can also act like giant, insulating blankets. The water droplets they're made of absorb and trap heat radiating from Earth, which helps to keep the planet warm.

$CO_2$ is mixed and distributed quickly throughout the Earth's

---

*Individual clouds differ widely in shape and size. Some stretch out over the horizon and pile up ten kilometers (6.2 miles) thick. Others are fleeting wisps. Some last ten minutes, others last an entire season.

atmosphere. Water vapor, in contrast, does not get very far before being transformed into clouds or fog. In fact, estimates of water vapor recycling are about ten days on average, whereas $CO_2$ is currently recycled about every one hundred to two hundred years! The effects of water vapor in the atmosphere as a greenhouse gas tend to be viewed as minimal in the context of global warming, because water vapor doesn't have time to accumulate in the same way as carbon dioxide, and as the water vapor forms clouds, it can even have a cooling effect. To be clear, water vapor is a powerful greenhouse gas, but the amount of water vapor in the atmosphere is closely tied to the amount of other greenhouse gases (especially $CO_2$). More $CO_2$ means more warming, which increases the amount of water vapor in the air. If the $CO_2$ goes away, so does the warming. Therefore, water vapor is not the culprit in global warming. Carbon dioxide is clearly the eight-hundred-pound gorilla in the room.

Additional indicators of human fingerprints on our changing climate are found in the layers in the atmosphere. There are five principal layers to Earth's atmosphere. Moving from the surface upward, these are the troposphere, stratosphere, mesosphere, thermosphere, and exosphere. The troposphere is where we (and most clouds) are all hanging out, and it is warming up over time (see Chapter 7). Perhaps surprisingly, the next layer up (the stratosphere) is actually cooling. This happens because the greenhouse gases are trapping the heat down in the troposphere like a blanket, preventing it from climbing into the upper layers.* If the sun were the cause of the warming, the entire atmosphere would heat up.

When things warm up, they expand, and when they cool, they shrink. If the troposphere is warming and the stratosphere is cooling, you'd expect the former to get bigger and the latter to contract. This is indeed what is happening. The height of the troposphere has increased several hundred meters since 1980. Traveling farther up

---

*Ozone losses from CFC emissions are an additional cause of stratospheric cooling—this is the part of the atmosphere where the ozone layer sits.

toward space, we find increasingly rarified air. For the same reasons as those for the stratospheric cooling, we would expect these layers to cool as a result of greenhouse warming (or warm as a result of increased solar activity). The thermosphere—the layer of the atmosphere where the International Space Station orbits—is indeed also cooling and shrinking.

So there you have it. No one has come up with a natural forcing or collection of forcings that can explain all of these observations. However, we do have a pretty simple explanation available, and it's one that meets Karl Popper's criteria: It is not only testable, but it has also been tested in innumerable studies. There are at least ten different indicators that we are to blame for the recent global warming trend. Everything we see points to our greenhouse-gas emissions as the cause. Occam's razor has shaved away everything the skeptics have come up with. It ain't natural. It's us.

# · 12 ·

# More carbon dioxide won't make a difference

**THIS SKEPTIC MYTH** offers a twist on denial, their usual approach. Here, Brad doesn't try to argue that the greenhouse effect is a bunch of malarkey, or that human carbon dioxide emissions aren't increasing the atmospheric concentration of $CO_2$, or even that climate change is a good thing. This myth claims that the $CO_2$ effect is saturated.

What Brad claims is that adding more carbon dioxide to the atmosphere won't make the greenhouse effect any stronger because the $CO_2$ that's already up there is absorbing all of the longwave (infrared) radiation coming up from the surface of the Earth. The greenhouse effect is caused by the sun heating up the planet's surface and clouds, primarily with light of shorter wavelengths (visible and ultraviolet). The warm surface then sends longwave radiation back up toward space, some of which is trapped by the greenhouse gases and sent back down to the surface, thereby functioning kind of like a blanket to keep the Earth at the reasonably comfortable temperatures that we're used to.

Think of it like insulation in your house. If you put in enough insulation, essentially no heat will escape out through your roof. Adding more insulation beyond that point wouldn't keep you any warmer, because there's no longer any heat getting out. Brad claims we're already at this point with carbon dioxide. So, in his view, we can emit all we want and suffer no further ill consequences. It's a free lunch.

Except that, famously, there is no such thing. Scientists have direct proof that the $CO_2$ effect is *not* saturated. This isn't even one of those "we're pretty sure" things that are rather common in the complex field of climate science. We're talking about straight-up proof. Before we can get to the evidence, though, we need to go over some fundamentals and a little backstory.

Molecules are not rigid things. The atoms that make up molecules wiggle and jiggle around all the time. Each atom can move in any direction. If you picture all the different ways in which the atoms within a molecule might be able to wiggle and jiggle with respect to one another, it seems awfully complicated. But all those vibrations are really just combinations of one or more simple motions. For example, in carbon dioxide, the two oxygen atoms are

## $CO_2$ MOLECULE VIBRATIONAL MODES

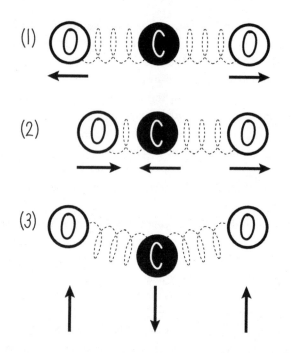

bound on opposite sides of a central carbon atom. All of the complex vibrational motions of $CO_2$ can be broken down into (1) a symmetric stretching motion, where the two oxygen atoms move closer and farther from the carbon in sync with each other, (2) an asymmetric stretching motion, where one oxygen atom moves closer to the carbon atom while the other oxygen atom moves farther away, and (3) a bending motion, where the molecule kinks back and forth in the middle.

Why are we telling you this? Well, it turns out that the energy of these vibrational motions in molecules corresponds to the amount of energy in longwave radiation. That means that when longwave radiation hits a molecule, the molecule can get excited and wiggle or jiggle even more. Each of the vibrations has a characteristic frequency, which means that only radiation having that same frequency (or a simple multiple of it) can make it go.* In addition to vibration, rotation of molecules is also a player in this game—and sometimes the vibrations and rotations even play together; think of it like crazy molecular crunking.†

Scientists routinely use this feature of molecules to identify what chemicals are present in a material. They shine infrared (longwave) radiation at it and measure how much of that radiation is successfully transmitted through. By varying the frequency (wavelength) of the radiation that they shine, they can figure out which frequencies the sample absorbs. Since every molecule has its own characteristic set of vibrational and rotational motions—and therefore frequencies—the scientist can deduce the chemical makeup of the sample.

One of the first scientists to apply these ideas to our atmosphere was an Irishman named John Tyndall. In the 1850s, he measured

---

*This is related to how microwave ovens work. Microwaves have a wavelength a bit longer than infrared radiation, and they can excite molecules like water or fat to rotate. Those molecules bump into their neighbors and transfer the energy all around. Moving molecules are what make something hot, so this heats up your food.

†Ask your kids.

the relative longwave absorptive powers of some of the gases that make up our atmosphere (nitrogen, oxygen, water, carbon dioxide, ozone, and methane). This absorption phenomenon is the physical mechanism of the greenhouse effect. The greenhouse effect had been discussed since at least 1824, when Joseph Fourier, a French mathematician and physicist,[*] proposed the idea, but Tyndall was the first to prove its existence with these measurements.[†]

It took about another forty years before the next notable event in the story behind this myth took place. In 1896, Swedish scientist Svante Arrhenius published a famous paper in *Philosophical Magazine* in which he calculated how changes in the amount of carbon dioxide in the atmosphere could influence the temperature of the Earth's surface through the greenhouse effect. This paper included his greenhouse law, which mathematically related the atmospheric $CO_2$ concentration to surface temperature. That relation, in modified form, is still in use today. Arrhenius was the first person to predict that $CO_2$ emissions from fossil fuels and other combustion processes were significant enough to trigger global warming.

Just a few years later, in 1900, fellow Swede and physicist Knut Ångström emerged as perhaps the first real skeptic of the concept of anthropogenic climate change. (You could say Knut is Brad's metaphysical great-grandfather.) He performed a relatively straightforward experiment, shining infrared (longwave) radiation through a tube containing carbon dioxide and measuring the absorption. He then removed some of the carbon dioxide from the tube and measured the absorption again, and he saw relatively little change in how much radiation was absorbed. Ångström's conclusion from this experiment was that it doesn't

---

[*]You might remember Fourier series from math class if you took it in college. Fourier actually developed these mathematical series to help him understand vibrations and heat transfer.

[†]Tyndall was also an early glaciologist! There are even several glaciers and mountains named after him because of his work in that field. Glacier melting wasn't a big deal back then, but thanks to climate disruption, it sure is now (see Chapter 8).

take very many $CO_2$ molecules to absorb lots of infrared radiation. He reasoned that adding more carbon dioxide molecules to the atmosphere would therefore not result in increased absorption of the longwave radiation emanating from the warm surface. In other words, he claimed that the $CO_2$ effect is saturated! It is this work, from more than a hundred years ago, that is the original basis for today's skeptic myth.

Arrhenius promptly disputed this conclusion in a paper published the following year. Interestingly, while Arrhenius does get credit as the first to note the connection between carbon emissions and global warming, he actually viewed the warming phenomenon as a positive thing, thinking it would foster greater agricultural productivity needed to feed a growing population. No one is right all the time! (See Chapter 3.)

Now here's a curveball to confuse you: It turns out that Knut's father was also a (much more famous) scientist, Anders Jonas Ångström. Anders was not only Knut's father but also one of the founding fathers of spectroscopy.* The field of spectroscopy is the study of interactions between radiant energy and matter—like, say, the interaction between longwave radiation and carbon dioxide molecules. A foundational tenet of spectroscopy is that the spectral absorption peaks of molecules are not perfectly sharp. This means that the absorption doesn't happen only at one precise frequency. Rather, a range of frequencies can be absorbed around the main (vibration) frequency while still exciting, say, a particular carbon-oxygen stretching motion.

There are several reasons for why molecules can absorb a range of frequencies like this. One big reason is something called Doppler broadening. You know how when an ambulance drives by you, the siren sounds high-pitched while it's approaching you and then low-pitched after it passes you? That's the Doppler effect. The sound

---

*His name has been immortalized in the atomic unit of measure called the Ångström, often abbreviated as Å.

waves get bunched up when the ambulance is getting closer to you and stretched apart when it's getting farther away, which changes the frequency of the sound. The same thing happens with molecules and longwave radiation. The molecules in the air are all bouncing around this way and that at high speed (like a bunch of ambulances in a crash derby, some moving closer and some farther away), which means that they can absorb (or emit) a range of frequencies, depending on their velocity.

As we'll explain shortly, Doppler broadening makes it harder to saturate the absorption of molecules in the atmosphere. There's a feature of Doppler broadening that makes it still more difficult to reach saturation. The broadening is very temperature-dependent (hotter molecules move faster, so there is more broadening of the absorption peak). So if temperature varies, as it does in different parts of the atmosphere, the longwave absorption peaks will have different widths at different altitudes in the atmosphere. This means that, when you look at the atmosphere as a whole, you see that the broadening effect is even broader. There are other sources of broadening, too, like pressure broadening (which also varies with altitude) and so-called natural broadening.[*]

Broad absorption peaks are one of the reasons that Knut Ångström was wrong when he concluded that just a tiny amount of carbon dioxide would absorb all of the longwave radiation. His simple experiment failed to capture the effects in the real atmosphere, because even when the central absorption frequency (the point right in the middle of the absorption peak) for a vibration in carbon dioxide nears saturation in the atmosphere, the fact that the peak widths are different at different altitudes means you can still have absorption of longwave radiation at frequencies on the wings of the saturated peak (that is, to the left and right of the central vibration frequency).

---

[*]Natural broadening is rooted in a quantum-mechanics concept called the uncertainty principle.

# ABSORPTION

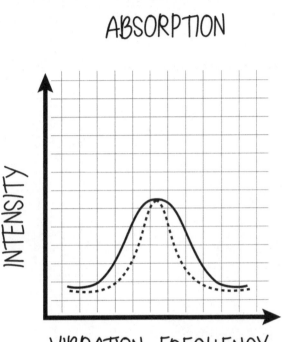

Schematic of a heavily broadened (solid line) and less broadened (dashed line) absorption peak. Here, "intensity" is referring to the amount of longwave radiation that the molecule absorbs at a given frequency.

Another reason that Knut Ångström's argument fails is that the atmosphere is not a static thing. Particularly in the troposphere—the region where the greenhouse gases are doing most of their work—there is a lot of convection, or moving air. The air rises and falls and gets blown all around. This movement of air mixes and distributes heat throughout that layer quite effectively. So when carbon dioxide molecules absorb the upward longwave radiation coming from the Earth's surface, they can transfer the resulting heat to other molecules in the air by bumping into them, enabling the $CO_2$ molecules to absorb more radiation. Maybe Knut should have paid more attention to his dad's lessons!

# CONVECTION

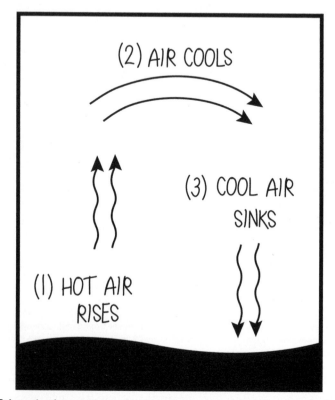

**Schematic of convection in the atmosphere, which thoroughly mixes the heat around, thereby enabling more of it to be absorbed than if the air were still.**

All right, so we've gone over the scientific principles underlying the greenhouse effect and why carbon dioxide's role in it is not likely to be saturated, as Brad claims it is. Now let's turn to the data. It should be a relatively simple thing to determine if the $CO_2$ effect is saturated or not.

You can measure the longwave radiation that escapes out into space by putting a satellite up there with a spectrometer to measure it. If the effect is saturated, then when we put more carbon dioxide into the atmosphere (by burning fossil fuels), there should be no change in the level of longwave radiation escaping. A corollary experiment would be to measure the longwave radiation bouncing back down from the atmosphere to the Earth's surface—the part of the radiation that's causing global warming. You can measure that with a spectrometer, too, without having to fuss with the whole launching-a-satellite thing. If the effect is saturated, this downward radiation should not be increasing with time, despite our continued $CO_2$ emissions. Let's look at both.

NASA put a satellite up in 1970 called the Infrared Interferometric Spectrometer, or IRIS. Other satellites with infrared spectrometers have followed (Interferometric Monitor for Greenhouse Gases, or IMG, launched by Japan in 1996; Atmospheric Infrared Sounder, or AIRS, launched by NASA in 2003; and the Aura satellite launched by NASA in 2004). Collectively, these space-based tools have collected data on longwave radiation escaping our atmosphere for over four decades. When you look at how the radiation making it through our atmosphere has changed over that time, you can clearly see that less longwave radiation is escaping at the frequencies where $CO_2$ absorbs today than was escaping in the recent past (the same is true for the powerful greenhouse gas methane). This is *direct evidence* that our carbon dioxide emissions are trapping more longwave radiation in the atmosphere. It is not saturated.

OK, so these gases are trapping more longwave radiation. What happens to that energy? It heats up the atmosphere. And what do hot things do? They emit longwave radiation, just like the Earth's surface does after it gets heated by the sun's rays. These longwave emissions from the atmosphere go in all directions. Some radiation works its way out into space (and is measured by those satellites), and some finds its way back down to the planet's surface.

Analogous to the satellite measurements, scientists have measured the downward longwave radiation from the 1970s to today, and they've found an unmistakable trend of increasing radiation over this period. Moreover, the increases are strongest at the frequencies tied to carbon dioxide (and methane). This is *direct evidence* that, beyond knowing just that our emissions are trapping more radiation, we know that our carbon dioxide emissions are producing a stronger greenhouse effect with time. It is not saturated!

Brad's argument is not just on thin ice at this point; it has fallen through. But there is one more piece to this puzzle. We've described

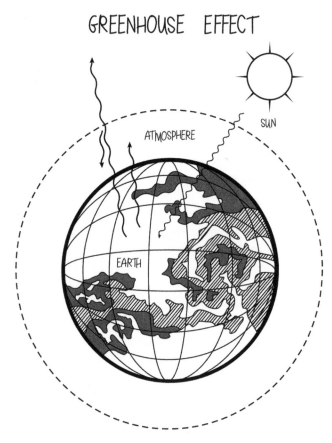

GREENHOUSE EFFECT

**A schematic of the greenhouse effect. The sun's rays heat up the Earth's surface, which then emits longwave radiation back toward space. Some of that radiation is absorbed by greenhouse gases, and some of that radiation is then reemitted back down to the surface.**

the proof that all those fossil fuels we're burning are strengthening the greenhouse effect on Earth, but how much does this greenhouse effect impact our climate? The term that scientists use to describe this correlation is the "climate sensitivity." This is the parameter that translates the greenhouse effect into global warming.

We can calculate the radiative forcing from greenhouse warming in watts per square meter (of the surface area of the Earth that is being warmed), expressed as $W/m^2$. If you multiply the radiative forcing by the climate sensitivity (units of degrees Celsius per watt per square meter), you get the temperature change expected from the forcing. The radiative forcing is getting stronger all the time because of the increasing carbon dioxide concentration in the atmosphere. Today it is about $1.7\ W/m^2$ (plus another $1\ W/m^2$ from other greenhouse gases). The climate sensitivity is roughly in the range of 0.5–1.2 degrees Celsius/$(W/m^2)$. So a good estimate for the temperature change expected from $CO_2$ greenhouse warming is 1.4 degrees Celsius.

The planet's surface has not yet heated up this much. The average temperature change since the Industrial Revolution is closer to 0.8 degrees Celsius. Where is the extra 0.6 degrees Celsius? It's in the oceans, which absorb a huge amount of the extra heat from global warming (see Chapter 7). It takes many, many years for that heat to evenly distribute itself through the whole system, including the oceans and the atmosphere. This means that, even if we completely stopped emitting carbon dioxide today, we would still see another 0.6 degrees Celsius of warming over time. This residual heat is called "warming in the pipeline." But of course, we aren't stopping these emissions today. In fact, our emissions worldwide are continuing to grow every year.

(Keep in mind that, while a few degrees doesn't sound like much, it makes a gigantic difference to the global climate. Check out Chapter 3 for a reminder of some of the anticipated [and current] impacts of climate disruption.)

Brad's done some math. He says "Fine, so our carbon emissions are still causing some additional warming, but all this talk of a runaway greenhouse effect is ridiculous. An exponential increase in $CO_2$ concentration in the atmosphere results only in a linear increase in the temperature, and that doesn't sound so frightening."

Brad's math is spot-on—this is indeed the relation between carbon dioxide concentration and temperature—but his Alfred E. Neuman–like "What, me worry?" attitude is misplaced. What Brad contends is that if, say, you double the concentration of carbon dioxide from pre-industrial times (that'd be from about 280 to about 560 parts per million), you'll have the same amount of temperature increase that you would from doubling it again from 560 to 1,120 ppm. On the surface, this seems like the temperature rise would slow down considerably over time. Let's dig under the surface a bit.

Assuming we follow a "business as usual" approach and don't make dramatic changes to our energy mix, conservative projections are that we'll reach that first doubling to 560 ppm by the year 2050, increasing to 900 ppm by the year 2100, and even more beyond that. The first doubling of $CO_2$ concentration would have taken about two centuries, and the second one only about seventy years. In other words, our business-as-usual scenario has us pumping so much carbon dioxide into the atmosphere that the temperature rise would accelerate (be faster than linear) despite the exponential/ linear relation between $CO_2$ and temperature.

What's more, $CO_2$-based greenhouse warming is just one forcing. As sea ice melts, more of the sun's energy will be absorbed by the planet. As Arctic permafrost peat bogs thaw, vast quantities of methane will be released into the atmosphere—and they'll be joined by further releases from melting methane clathrates* in the ocean. (Remember that methane is a greenhouse gas more than

---

*Methane clathrates are solid compounds in which methane is trapped within a crystal structure of water, forming a solid similar to ice. Significant deposits of methane clathrate exist under sediments on the ocean floor.

twenty times stronger than carbon dioxide!) Rainforests in some regions will release their stored carbon as they dry out. Same goes for broad swaths of land that will experience desertification. More water vapor will enter the atmosphere due to increased evaporation, and remember that water vapor is a potent greenhouse gas itself, thereby warming things further. These are examples of positive feedback cycles that will amplify warming, accelerating it beyond the already accelerating warming from our own emissions.

Yes, Brad, me worry.

## · 13 ·

# What about . . .

THIS CHAPTER IS a catch-all for a collection of various myths, some seemingly reasonable at a cursory glance and some rather far-fetched, but all put forward by the climate-skeptic community. One thing that Brad and friends don't lack is creativity. In many cases, the arguments against these myths are quite simple, but because they are sometimes from so far out in left field, odds are you haven't considered them before. We will systematically knock down each of these often-wild theories with the facts. It is important to keep in mind that, above and beyond the facts we'll discuss in this chapter, none of the myths presented here can provide an alternate explanation for all of the phenomena we've linked to anthropogenic climate change in the previous chapters. The *only* explanation that fits all the data is that our greenhouse-gas emissions are causing the climate disruption that we're all experiencing and that will continue to worsen in the future.

## VOLCANOES

Humans aren't the only things adding carbon dioxide to the atmosphere. Skeptics will argue that volcanic eruptions release an amount of $CO_2$ that dwarfs the human contribution.

What are the numbers? Volcanoes release material both on land and under the oceans. Scientists estimate the $CO_2$ emissions from the former at about 260 million (with an "m") tons per year and emissions from the latter a somewhat more modest 90 million tons per year. (Of course, the actual activity can vary a lot from year to year, depending on the scale and number of eruptions.)

Almost all of the carbon dioxide produced by the underwater volcanoes is absorbed by carbon sinks in the ocean, with a good example of such a sink being the fresh lava that is formed as part of the same volcanic activity*—so little makes its way into the atmosphere. But even if all of the $CO_2$ emissions from both the underwater and land-based volcanoes added to the atmospheric content, on average the total would be about 350 million (with an "m") tons of carbon dioxide added to the atmosphere each year. Sounds like a lot of $CO_2$, but human $CO_2$ emissions are roughly 30 billion (with a "b") tons per year—and rising. This means that humans are emitting *at least eighty-five times* more carbon dioxide than the volcanoes.

Thus the contribution from volcanoes is minimal. In fact, this is clear from direct measurements of $CO_2$ in the atmosphere over the past fifty years. If volcanoes were a major contributor, there would be spikes in the carbon dioxide concentration following major eruptions like the Agung eruption in 1963, the Mount St. Helens eruption in 1980, the El Chichón eruption in 1982, and the Pinatubo eruption in 1991. However, the $CO_2$ measurements show no rise whatsoever associated with these events. These volcanic emissions are simply too small to make a difference.[†]

## PACIFIC DECADAL OSCILLATION

The Pacific Decadal Oscillation (PDO) is a climate phenomenon akin to the Southern Oscillation that leads to El Niño/La Niña

---

*Lava is composed of minerals such as olivine and feldspar, which can combine with carbon dioxide to form solid carbonate materials.

[†]Even so, sulfate aerosols blasted into the atmosphere from volcanic eruptions do have a significant, though relatively short-lived, cooling impact on climate.

cycles. The PDO is less regular than the Southern Oscillation, and the shifts from warm phases to cool phases typically take anywhere from a few years to forty years.

Skeptics like to point out that the PDO was in a warm phase during the early part of the twentieth century (when global temperatures rose), was in a cool phase during the middle of the twentieth century (when global temperatures cooled), and returned to a warm phase in the latter part of the twentieth century (when global temperatures have risen again). This apparent correlation might lead you to believe that the PDO is what's actually driving the climate changes over this period, rather than human carbon dioxide emissions. This is a popular myth among skeptics because of its allure as a seemingly good alternative explanation for temperature trends in the twentieth century.

The clearest evidence refuting this argument is apparent when the average global temperature and the PDO index (how much the North Pacific sea-surface temperature is above or below normal) are plotted over the entire twentieth century. The temperature plot has an obvious upward trend (see Chapter 7), while the PDO index is flat. This result is not surprising. PDO is an example of something called internal variability. It involves movement of heat from one part of the Earth to another, shifting back and forth between the oceans and the air. There is no net change in the amount of heat in the system. Global warming, in contrast, involves an overall addition of heat to the Earth—heat from the greenhouse effect strengthened by our fossil-fuel use (see Chapter 10).

An analogy would be if you had two glasses and some water. You could pour the water back and forth between these two glasses to your heart's content, making one glass or the other have more or less water. That's internal variability. Global warming is adding more water from the faucet. Internal variations like the PDO cannot explain a long-term temperature trend. Over the long run, both the oceans and the air are getting warmer. That has nothing to do with the PDO.

## COSMIC RAYS

First things first. What the heck is a cosmic ray? Well, it isn't actually a ray at all. Cosmic rays are very high-energy particles (usually atomic nuclei) screaming through the universe like a Lamborghini in a wide-open carpool lane. And by very high energy, we mean *very* high energy. We're talking about particles with energies tens of millions of times higher than what scientists can produce in the most powerful atom smasher on the planet (the Large Hadron Collider). It's a little weird to think about it, but cosmic rays are actually passing right through you as you're reading this.

They play various important roles on Earth—some helpful and some not, such as creating carbon-14 (the radioactive isotope used for dating ancient items), causing damage to electronics and life-forms (yes, including you), and helping to trigger lightning strikes. So what do these speed demons have to do with climate change? Nothing, but skeptics have argued otherwise.

This myth, which first surfaced in the late 1950s and is repeated to this day, goes like this: The strength of the sun's magnetic field is increasing over time, and therefore the field is deflecting an increasing number of cosmic rays away from the Earth, with the result that fewer low-level clouds form in the atmosphere, leading to the observed global warming (this type of cloud has a cooling effect, so fewer clouds would mean more warming). For the argument to be valid, each aspect of the logic must hold. That means that the sun's magnetic field must really have been increasing for decades, cosmic-ray counts on Earth must have been dropping over that period, and, importantly, there must be a connection between cosmic rays and cloud formation. Also, low-level cloud cover should have been decreasing for decades, if this is the real explanation for global warming. We have data for most of these aspects, so let's see what the real story is.

As we discussed in Chapter 11, the sun does go through various changes in its activity—including its magnetic activity. Instead of increasing, over recent decades the magnetic activity has not increased at all. It's actually decreased a little. That alone scuttles this myth, but we'll go through the rest (we can't help ourselves).

Scientists have been measuring the number of cosmic rays on Earth since about 1950. The particle counts vary from year to year, but there is no apparent long-term trend up or down over this time period. Most recently (since about 1990), the rate of cosmic-ray hits has actually been on an upswing—the opposite of the direction purported by this myth. This measurement, too, shows this skeptic argument doesn't hold water, but let's keep going.

For this myth to have any credibility, cosmic rays would have to be able to influence the number of low-level clouds on Earth. The supposed mechanism for this influence is that (1) cosmic rays stimulate the formation of aerosols, (2) the aerosols grow through condensation of gases in the air to form something called cloud-condensation nuclei (basically, little particles that enable water to change from gas to liquid), and (3) the cloud-condensation nuclei cause increased cloud formation. Step 1 in this process does indeed happen. Step 2 is where things seem to break down. A number of studies in recent years have found little or no connection between cosmic rays and cloud-condensation nuclei or subsequent cloud formation. Once more, this skeptic argument simply isn't supported by the data.

One of the reasons that this myth, like Grigori Rasputin, just won't die is that there was a period from about 1984 to 1991 during which cloud cover did seem to track the cosmic-ray count data. This is a classic case of skeptics cherry-picking a connection that was most likely coincidence. The seeming correlation between cosmic rays and clouds diverges completely after 1991 (and before 1984), so if you look at the full set of data, it's clear that there is no connection there. Perhaps even more importantly, there is also no correlation between cosmic-ray counts and the global average temperature over the past four decades. This myth is busted.

## SOOT

Mankind spews all sorts of pollutants into the air—not just carbon dioxide. Some skeptics have claimed that a big part of global warming is attributable to soot rather than $CO_2$. Soot comes out of the exhaust pipes of many diesel engines and from burning coal, wood, and dung.* The technical term for soot is "black carbon," and it does in fact contribute to global warming. One way that black carbon warms up the planet is by directly absorbing sunlight when the carbon is up in the air as little particles. Just like when you get hotter wearing black clothing in the sun than while wearing white clothing, these black particles soak up lots of radiant energy and then heat up the air around them. The key is to note that black-carbon particles don't stay in the air very long. The second way that black carbon warms the planet is when these particles fall out of the sky and land on snow or ice. This makes the ground much less reflective and, therefore, warmer. This phenomenon leads to faster melting of ice in glaciers and the like.

So black carbon is a contributor to global warming, but how big is that contribution compared to that from carbon dioxide and its greenhouse warming? Projections are that it should be possible to achieve 0.1–0.2 degrees Celsius less warming between now and 2050 if we dramatically reduce black-carbon emissions. That's a good thing to do—and not just because of the decreased warming; black carbon is also linked to a number of serious health effects like coronary heart disease—but the warming that we're all concerned about is more than ten times as many degrees Celsius. Soot is a minor player and cannot explain the warming we've witnessed over the past century.

---

*Exhaust pipes and coal-fired power plants usually have filters in their smokestacks to reduce their soot emissions.

## GLOBAL BRIGHTENING

Global brightening (and global dimming) refers to changes in the amount of sunlight making its way down to the Earth's surface. Brightening means there is a gradual increase in the amount of direct irradiance, which is what scientists have observed over the past few decades. Though the level jumps around quite a bit, there has been a roughly linear increase of 0.16 $W/m^2$ per year since the 1980s. More sunlight hitting the surface should warm things up, so skeptics claim that this global brightening is the reason we've witnessed global warming over the same period. This sounds like a reasonable argument—until you look at the reasons why global brightening has been happening and realize just how complex the story actually is.

The three main drivers of brightening have been reductions in the atmosphere of cloud cover, sulfate aerosols, and soot. (The reduction in sulfate aerosols and soot are the result of pollution-prevention regulations around the world, such as the Clean Air Act.) Obviously, clouds block out the sunlight, so fewer clouds means more light reaches the surface. Sulfate aerosols scatter sunlight and soot absorbs it (as we just discussed), so reductions in those, too, will result in global brightening. However, the influence on climate of each of these changes is non-trivial.

Aerosols remain the forcing with the greatest uncertainty in climate change because they have a multitude of effects on the temperature, some direct (like scattering sunlight, a cooling effect) and some indirect (like influencing cloud formation, potentially either cooling or warming). Clouds may block sunlight from reaching the surface (a cooling effect), but they also act as traps for longwave radiation similar to greenhouse gases (a warming effect). Soot can prevent sunlight from making it down to the surface (a cooling effect) but also absorbs sunlight in the atmosphere (a warming effect). If you're confused at this point, you're not alone. The point is that it isn't clear how large the influence of global brightening is on global temperatures—not even whether it's principally a warming or cooling influence. In contrast, there is little doubt that greenhouse warming from carbon dioxide emissions is a major influence for warming and is the primary driver of recent climate disruption.

## OTHER PLANETS ARE WARMING

This myth isn't so much out of left field as out of this world—literally. It has become popular for climate skeptics to claim that other planets in our solar system are warming, which must mean that the sun is the culprit, since obviously there aren't any humans out there (and Marvin the Martian's Illudium Q-36 Explosive Space Modulator presumably doesn't have a large carbon footprint). We described why the sun can't explain warming on Earth in Chapter 11, and the same evidence applies to other planets as well. The sun's radiant output has not been increasing over recent decades, so it is not the cause of our warming. But are other planets even warming in the first place?

It turns out that this question is not so easy to answer, and the evidence that skeptics point to for their warming is, as is so often the case, cherry-picked. Think about how challenging it is for scientists to determine the temperature trends here on Earth, where we have really good instruments that have been collecting data from thousands of locations around the planet for many years. The situation on other planets is significantly different. The "evidence" that Mars is warming comes from just two pictures, one taken in 1977 by the Viking spacecraft and one taken in 1999 by the Mars Global Surveyor. The surface of Mars in the 1977 photograph was brighter than in 1999, which a mistaken NASA scientist attributed to a difference in temperature caused by surface dust changing how much light Mars absorbs. Here is a classic climate-vs.-weather error. Weather is what's happening at a particular moment, whereas climate is weather averaged over many years. You can't take single days separated by twenty-two years and say much of anything about climate. Regardless, the sun, which in this myth is the sole cause of warming, clearly has nothing to do with it.

Skeptics also point to planets like Neptune and its largest moon, Triton, as supposedly warming.* They have brightened over time, which skeptics claim means that they're warming due to increased output from the sun. The truth is that Neptune and Triton are brightening because of seasonal changes, albeit extremely slow ones. If you've ever felt like winter was lasting forever, be glad you don't live on Neptune, where it lasts about 41 years! A full orbit of Neptune around the sun takes 164.8 Earth years. Since the planet was observed for the first time in 1846, and the brightening data was collected far more recently, we obviously don't have anywhere near enough information to determine if Neptune is experiencing any particular long-term (multi-Neptune-year) temperature trend.

## OZONE

Ozone, a gas composed of three oxygen atoms, makes up a small percentage of our atmosphere, but it has an important role to play. Most important, it absorbs lots of ultraviolet radiation, protecting us from skin cancer and other health problems. When it absorbs that UV radiation, it prevents the radiation from reaching the Earth's surface and warming things up down here. Skeptics claim that, because ozone levels are falling, we expect more UV light to reach the surface, which would explain our observations of global warming.

This one is easy to refute. Ozone levels aren't actually dropping anymore. They were falling previously because of human emissions of chlorofluorocarbons (CFCs . . . see the next myth for more on these) from dry cleaning, refrigerants, and plenty of hairspray in

---

*Pluto is another one, but we'll leave that poor celestial body alone, since it's still recovering from being demoted from a planet to a dwarf planet.

the 1980s. But regulations passed in the 1990s that limited CFC use led to an abrupt halt to ozone depletion. Ozone levels have actually been climbing back up since that time. Changes in ozone levels don't explain global warming.

## CHLOROFLUOROCARBONS (CFCS)

A paper written by Qing-Bin Lu and published in a little-known physics journal called the *International Journal of Modern Physics B* has captured the attention of skeptics recently. Among other things, Lu attempts to make a case that CFCs, rather than $CO_2$, are the primary driver of climate change. There is plenty of evidence that $CO_2$ is to blame (see Chapter 10), but let's see what Lu has to say.

He examined global-surface temperatures in the first decade of the twenty-first century and compared them to both CFC and $CO_2$ levels in the atmosphere. Over that time, CFC levels and temperatures have been approximately steady, whereas $CO_2$ levels have continued to rise. He concludes that CFCs fit the temperature data better and are therefore the cause of global warming. There are numerous errors in his analysis (one being that ten years is not enough time to track a climate trend—thirty years is the standard), but the key one for this discussion is that surface temperatures are not a good measure of the overall heat on the planet. In Chapter 7, we talked about how the oceans absorb the vast majority of the extra heat from global warming, and they have indeed been soaking up unfathomably large amounts of heat in recent years. CFCs are, in fact, powerful greenhouse gases, but the number of CFCs in the atmosphere is so trifling that their warming effect is small in comparison with that of $CO_2$.

## WASTE HEAT

Waste heat is generated essentially every time energy is used. Think of the hot engine after you drive your car or your laptop burning your legs. On a larger scale, this happens in power plants, too, whether they be coal-fired, natural-gas, or nuclear. All that waste heat ends up being distributed throughout the environment, which makes temperatures rise.* Skeptics have claimed that waste heat represents a far larger contribution to global warming than does the greenhouse effect.

If you run the numbers, this myth is laughable. Globally, waste heat accounts for about 0.028 $W/m^2$ of warming. Greenhouse warming, from $CO_2$ and other gases combined, is about 2.9 $W/m^2$— that's one hundred times larger than the waste-heat contribution.

## SATELLITE MICROWAVE TRANSMISSIONS

This is one of those crazy myths. Some skeptics have argued that the time frame of global warming has coincided with our use of satellites orbiting the Earth and transmitting microwave signals back and forth. As you know from the microwave oven in your kitchen, microwaves passing through water and many other materials makes them heat up. This myth argues that this same mechanism is heating up the planet. They claim that microwave satellite transmissions are warming up our atmosphere!

Your microwave oven operates at around 1,000 watts. Satellites transmit with an average power of about 3,000 watts, though a few operate closer to 20,000 watts. There are fewer than one thousand satellites orbiting the Earth today. If you add up all the power from

---

*Waste heat from power plants is another reason to shift away from fossil fuels (and nuclear, for that matter). Thermal pollution depletes oxygen levels in water, which wrecks ecosystems in the vicinity of power plants.

these satellites, assuming they are operating nonstop and directing all of their power at the Earth (neither of which is typically the case), you still end up with under 3 megawatts (MW) of microwave power in total. Spread out over the surface of the planet, that equates to about 0.0000005 W/m².

That's over a million times smaller than the effect of greenhouse warming. If the waste-heat myth is laughable, this one qualifies as ludicrous.

## SECOND LAW OF THERMODYNAMICS

There are four fundamental laws of thermodynamics: the first, second, third, and . . . zeroth (!?). (Sounds like someone started numbering things prematurely and had to stick one in there later, like Harry Potter's Platform 9¾.) Putting aside scientists' seeming inability to count like normal people, here's a quick summary of the four laws (don't worry, there won't be a quiz later):

1. If two systems are both in thermal equilibrium with a third system, then they are in thermal equilibrium with each other.
2. The increase in internal energy of a body is equal to the heat supplied to the body minus work done by the body. (This involves the principle of conservation of energy.)
3. When two initially isolated systems interact, the entropy of the isolated systems is less than or equal to the total entropy of the final combination.*
4. The entropy of a system approaches a constant value as the temperature approaches zero.

---

*Entropy is a measure of the number of specific ways in which a system may be arranged, often viewed as a measure of disorder. For example, once you've stirred cream into your coffee, you can't un-stir it back out.

Rudolph Clausius, a nineteenth-century German physicist and a founding father of thermodynamics as a science, stated the second law as follows: *Heat cannot flow spontaneously from a material at lower temperature to a material at higher temperature.* In other words, if you place an ice cube in your coffee, the ice cube isn't going to get colder nor the coffee hotter.

OK, so how does this connect with climate change? Skeptics argue that the greenhouse gases in the atmosphere are cooler than the Earth's surface, so surely they cannot warm the planet. In their view, the greenhouse effect itself would represent a violation of the second law of thermodynamics. Recall from Chapter 12 that we have direct evidence that the greenhouse effect is indeed occurring, and rest assured that the laws of thermodynamics have not been broken. So where did the skeptics err?

You can think of greenhouse gases as a blanket wrapping the Earth. A blanket doesn't generate its own heat, but it does help keep you warm. What it's doing is trapping some of the heat your body makes, rather than letting it escape to the environment around you. Putting a blanket on, say, a rock doesn't make the rock warmer— the rock isn't generating any heat to be trapped. So the greenhouse effect is perfectly consistent with the second law of thermodynamics (and all the others, too, even that weird zeroth one).

. . .

As you can see, skeptics come up with quite an assortment of arguments in an attempt to draw attention away from the fact that human activity is driving the disruption of our climate. It's healthy to question any explanation, whether it be soot, cosmic rays, or human greenhouse-gas emissions—and to compare that explanation with the full breadth of data that we have available. For an explanation of climate change to be valid, it would have to at least be consistent with these data. None of the skeptic arguments passes this test. The only explanation available that's consistent with the range of observations across the globe and over time is that we are the primary drivers of climate disruption through our emissions of carbon dioxide.

**5**

# There's Nothing We Can Do About It

Carbon taxes or cap-and-trade systems will destroy the economy, kill jobs, and hurt the poor

Renewable energy is too expensive or too variable

# 14

# Carbon taxes or cap-and-trade systems will destroy the economy, kill jobs, and hurt the poor

**IF WE ACKNOWLEDGE** that carbon emissions are fueling climate disruption (Chapters 1 and 10) and that its impacts will be substantially negative to society as a whole (Chapter 3), we should explore possible solutions. The most common policy solution put forward is to monetize the hidden costs associated with fossil fuels through either a carbon tax or a cap-and-trade system. If these costs are put in front of the consumer in a straightforward way, the economic decision to switch from fossil-fuel sources to energy sources with a lower carbon footprint is clear. Yet skeptics argue that the cost of these policies would cripple the economy and disproportionately hurt the poor. In this chapter, we'll work our way from exploring the hidden costs of traditional energy, to the proposed policy approaches, to the projected impact of these policies.

Energy use has an apparent cost. You pay at the gas station when you fill up your car, and you pay electric and gas bills to your utility providers every month to heat, cool, and power your home. While those payments may seem burdensome at times, you are actually paying far more for the use of the energy than the bills show. You aren't paying most of the costs of energy at the pump or to your utility company. These other costs are hidden. Economists refer to them as "external" costs, and they can wreak havoc economically if you don't account for them properly.

Just a few examples of the countless external costs related to fossil fuels are:

- Higher health-care costs for asthma attacks, bronchitis, and other cardiopulmonary problems caused by particulate, nitrogen oxide, and ozone emissions from coal-fired power plants
- Heavy metals like mercury released by burning coal, that enter our food system and cause poisoning and developmental disorders associated with health care and other costs
- Oil spills, natural gas explosions, and coal mine collapses leading to insurance (home, health, business, and—most unfortunately—life) and government assistance (tax) costs
- Nitrogen oxide and sulfur dioxide emissions that lead to acid-rain damage to infrastructure (buildings, bridges, and so on) and crops, resulting in replacement/repair costs and higher food prices
- Higher health care and water costs because of coal-ash pollution of freshwater resources
- National security costs to secure overseas fossil fuel resources
- Tax breaks for fossil fuels
- And, larger than all of the above combined, myriad costs related to climate disruption and ocean acidification (see Chapter 3)

Calculating exactly how big these external costs are is an enormous challenge, but many economists have given it a go. The expression used to capture these costs is the "social cost of carbon" (or SCC), and it is usually represented in dollars per ton of $CO_2$ emitted. Given the large uncertainties involved in this sort of calculation, values for the SCC vary considerably, but a typical range from various studies is about 50–260 dollars/ton $CO_2$. Recall that annual global emissions of carbon dioxide are measured in tens of billions of tons, so we're talking about some colossal external costs here. We are all paying this SCC, whether we know it or not, and whether we like it or not, and the only way to minimize these costs is to use less fossil fuel.

Some skeptics argue that limiting carbon emissions is a pointless exercise because the climate system is very slow to react, which would imply that warming and sea-level rise will continue for centuries even if $CO_2$ emissions were halted today. While it's true that it takes hundreds of years for the climate system to equilibrate, continuing our business-as-usual approach with fossil-fuel use would dramatically exacerbate the problems we've already initiated, making the changes happen far sooner and with far greater severity. Closing your eyes doesn't make your problems disappear. The hope is that, by buying ourselves some time through limits on emissions now, we will be able to develop effective technologies to pull excess $CO_2$ out of the atmosphere (through carbon capture and sequestration), thereby avoiding the worst consequences of climate disruption in the future.

Perhaps the biggest challenge in convincing countries to put limits on their carbon emissions (other than people like Brad doing their best to say it isn't necessary) is a phenomenon known as the tragedy of the commons. In economics, the tragedy of the commons is the depletion of a common resource by individuals, all acting rationally and independently according to their own self-interest, despite their understanding that using up the shared resource is not in their long-term best interest. Some examples of the tragedy of the commons are overfishing (each fisherman chooses to benefit in the short-term by fishing as much as he can, and in the end there aren't enough fish for anybody), use of freeways (each driver opts to drive on the freeway because it's supposed to be the fastest route, but then everyone ends up stuck in traffic because there are too many cars), and careless use of public restrooms (let's not go into detail on this one . . .).

In the context of climate change, the tragedy of the commons is that if each country looks out only for its own best, short-term interests, it will opt not to enact limits on carbon emissions because there is a short-term cost for doing so. The way we can overcome the tragedy of the commons is to act collectively. If all countries act together, we achieve the best long-term outcome for everyone. This is what international climate conferences are all about.

The Kyoto Protocol, a treaty adopted in 1997 that sets binding obligations on industrialized countries to reduce emissions of greenhouse gases, was the first attempt to act collectively on an international scale. Unfortunately, countries like the United States and Canada didn't sign on the dotted line, which greatly reduces the effectiveness of this agreement.* This is the sort of thing that makes the tragedy of commons so, well, tragic.

Since Kyoto didn't get the job done, new efforts are needed. So how does a country, or a group of countries, limit carbon emissions? The two most common mechanisms bandied about are cap-and-trade systems and carbon taxes. Let's briefly go over what each of these entails.

Cap-and-trade, also called emissions trading, uses economic incentives to reduce pollution. A government (or other central authority) sets a limit on the total amount of emissions for, say, carbon dioxide. This "cap" is split up into chunks and allocated or sold to companies in the form of emissions permits, which represent the right to emit a certain number of tons of $CO_2$ per year. If a company needs to increase its emissions, it has to buy permits from others who have permits to spare. (This is the "trade" part of cap-and-trade.) The company buying the permits is effectively paying a charge for polluting, while the seller is rewarded for having reduced its emissions. This is a market-based approach, and the idea is that the companies that can reduce emissions at the lowest cost will do so. Ideally, such an approach accomplishes the desired carbon-emission reductions at the lowest overall cost to society.

A carbon tax is a more straightforward, transparent way to capture the external costs associated with fossil fuels. In this approach,

---

*Actually, Canada initially agreed to the Kyoto Protocol, but then it withdrew in 2012. The United States never signed on at all.

a government simply puts a price on carbon emissions. The object is supposed to be monetizing the SCC that we discussed earlier in the chapter. In other words, it's about having everyone pay the real cost for what they're buying from the seller rather than unknowingly handing part of the cost to others. Not having a carbon tax is like going to the store to buy a television that's listed at four hundred dollars but really costs eight hundred dollars; you give four hundred dollars to the store, then a while later have to give three hundred dollars to the gas station down the block and one hundred dollars to your doctor—doesn't make much sense, does it?

There is some dollar amount per ton of $CO_2$ emitted that would represent the actual external (hidden) cost associated with adding that much pollution to the atmosphere. In practice, the carbon taxes that have been proposed in places like the US Congress (but never passed into law) and those that have been enacted for certain fossil-fuel sources elsewhere (including in Japan, India, and many countries in the European Union) are far below the estimated SCC. Rather than 50–260 dollars/ton $CO_2$, these taxes tend to be closer to 10–30 dollars/ton $CO_2$. Even though these taxes are low-balling the true costs, climate skeptics still argue that implementing a carbon tax will dramatically harm the economy. Economists with expertise in climate disagree, with a consensus of greater than 95 percent arguing that government action on climate change is needed.[*] Interestingly, this level of consensus is comparable to that of climate scientists' consensus that humans are the primary driver of climate disruption (roughly 97 percent; see Chapter 1).

A crucial mistake made in these skeptic arguments is that they focus only on the cost related to pricing carbon and completely ignore all the benefits. It's not much of a cost-benefit analysis when you skip the benefit part! It doesn't take an economist to tell you that a tax generates a revenue stream for the government. That

---

[*]From a survey entitled *Economists and Climate Change: Consensus and Open Questions*, conducted in 2009 by the Institute for Policy Integrity at the New York University School of Law.

money doesn't just disappear (no doubt some of you are scoffing at that thought). There are various ways that revenue can be distributed. One approach is to lower other taxes, which would make the carbon tax "revenue neutral." More proactive approaches might entail funding energy-efficiency programs (like subsidies for homeowners to improve insulation or adopt more efficient appliances) or funding research on more sustainable energy technologies. This type of investment offsets much of the cost of the carbon tax.* Above and beyond this, though, is the enormous benefit of a reduction in carbon emissions, which will ultimately save many trillions of dollars globally.

For argument's sake, we'll follow the skeptics and ignore the benefits. Let's take a closer look at the potential impact just from the cost side.

Most of the skeptic attention has been directed at US policy, so we'll stick with the domestic emission-reduction laws that have been proposed in recent years (examples include the Climate Security Act of 2008 introduced by senators Joe Lieberman and Mark Warner and the American Clean Energy and Security Act of 2009 introduced by Congressmen Henry Waxman and Ed Markey). A number of nonpartisan organizations, including the Congressional Budget Office (CBO), Energy Information Administration, and Peterson Institute for International Economics, have evaluated the costs of these proposals. Nearly every one of these analyses projects a cost to the US economy of well below 1 percent of GDP. Even ignoring the benefits, this is a far cry from the destruction of the economy that the skeptics claim will result.

The impact on household costs would be less than one dollar per person per week—mostly from increased gasoline prices. (Usually these bills include provisions to shield low-income house-

---

*Google.org performed an analysis taking some of these benefits into account (specifically, the value of clean energy innovation driven by a thirty dollars/ton price on utility-sector carbon emissions), and they projected a strong, sustained *increase* in GDP.

holds from the costs.) Utility bills would go up by a couple bucks a month, though if energy-efficiency programs were implemented, these bills would most likely decrease rather than rise.

You can gain a broader perspective by looking at projected impacts at the national scale. CBO analysis of these bills has consistently projected a *decrease*, measured in billions of dollars, in the federal deficit. Another strategic advantage comes from increased energy independence. While the United States will eventually be largely independent from foreign oil as a result of new technologies for extraction of domestic supplies (probably in the 2030–2050 time frame), a carbon tax is projected to spur the country to dramatically reduce its consumption of (foreign) oil as soon as it's enacted.

Even if researchers are way off in their estimates for the SCC (as skeptics claim but don't demonstrate), a SCC value as low as ten dollars/ton $CO_2$ would mean that the direct benefits of the carbon taxes being discussed would be enough to completely offset the cost. Since the SCC is almost certainly much higher than this number, it is clear that policy-based mitigation of climate change is several times less costly than trying to adapt to the disruption once it's happening.

A related skeptic claim is that carbon taxes or cap-and-trade systems will kill jobs, a topic that resonates with populations suffering from high unemployment. An illustration of such a claim comes from Gabriel Calzada Álvarez, the founder and president of the Fundacion Juan de Mariana, a libertarian think tank based in Spain. He attests that solar-energy investments destroyed 15,000 jobs over the course of a year. Calzada is cherry-picking the data just like skeptics do with climate data (in fact, Calzada has been a featured speaker at global warming science denial conferences). We should note that both the US and Spanish governments (and *The Wall Street Journal*) have pointed to numerous factual and methodological errors in Calzada's analysis. The Spanish Ministry of Labor has found that, contrary to Calzada's study, renewable-energy industries have created 175,000 jobs. In the big picture, far more jobs were created than were lost.

Generally speaking, investments in renewable energy create more jobs than those in fossil energy because renewables require a greater share of manufacturing, installation, and maintenance labor than what's involved in extracting and transporting fossil fuels. The Regional Greenhouse Gas Initiative, for example, involved ten northeastern states implementing a cap-and-trade system for the utility-power sector. A 2011 study of the effectiveness of the program after its first three years found that the system had added 1.6 billion dollars in value to the economies of the participating states (far more than it cost) and had created 16,000 jobs.[*] Limits on carbon emissions don't kill jobs. They create them.[†]

So we've established that limiting carbon emissions will cost little, likely increase GDP, reduce government deficits, increase energy security, and create jobs. But what about the poor? Will low-income families suffer the costs of these climate-change mitigation programs disproportionately?

There are two ways to answer this question. The first is that nearly all proposed or enacted legislation related to cap-and-trade or carbon taxes includes language that offsets the household costs incurred by low-income families. These offsets are often in the form of direct subsidies of some homeowners' electric bills. As an example, CBO analysis of the Waxman-Markey bill determined that low-income households would actually experience a decrease in their annual costs if the bill were to become law.

---

[*]Analysis Group, Inc., *The Economic Impacts of the Regional Greenhouse Gas Initiative on Ten Northeast and Mid-Atlantic States*, 2011.

[†]Of course, if you're a coal miner, you may indeed lose your job. This impact should not be ignored or whitewashed, but the societal benefits so outweigh such consequences that the choice is clear. Moreover, there are ways in which these types of localized job losses could be mitigated, such as through education and job-training programs to prepare those workers negatively impacted for new jobs funded with resources collected through a carbon tax. Moving on from a career associated with significant health and safety risks (mine collapses/explosions/floods, black lung disease, and so on) would be yet a further advantage.

But there's a bigger picture here, and that's the second way to answer this question. A 2011 study led James Samson from McGill University and published in *Global Ecology and Biogeography* examined the issue of where carbon emissions come from in comparison to which regions of the world are most vulnerable to climate disruption. Their global maps of these two parameters show an alarming contrast: Those regions of the planet most vulnerable (Africa, the Middle East, Southeast Asia, and South America) are also the regions with the smallest carbon emissions.* In other words, the developed world's actions are disproportionately impacting the developing world. If the poor are your concern, limiting carbon emissions should be among your top priorities.

Folks are getting the message. The World Bank was founded after the Second World War to combat global poverty, and mitigating climate disruption has become its new guiding principle. In a study by the World Bank in 2013, climate change was projected to cause African countries to lose up to 80 percent of their cropland by the 2030s. The same study projected that large portions of Thailand and Vietnam would be flooded. The World Bank is therefore focusing its efforts on policies and practices that can reduce fossil-fuel use. When the world's foremost poverty-fighting organization puts climate change at the very top of its concerns, it tells you how much credence you should put on the skeptic claims about mitigation strategies hurting the poor.

---

*There are more than a billion people who do not have access to modern energy today, mostly living in poor countries. Bringing energy to them, even if we did so with coal (which is prevalent in many developing countries), would constitute less than 1 percent of worldwide emissions.

# 15

# Renewable energy is too expensive or too variable

**WE STARTED THIS** book with a look at where our energy comes from today. Worldwide, we use energy at a rate of about 18 terawatts (TW).* That number is projected to reach about 25 TW by the year 2035, growing further to 30 TW by 2050—and still more beyond that. The vast majority of our current 18 TW comes from fossil fuels, with nuclear energy representing about 6 percent and renewable sources the remaining 10 percent. Breaking down the renewable fraction further, we find that about three-quarters of it comes from hydroelectric power. In other words, despite all the media attention to wind, geothermal, solar, and other (non-hydro) renewable energy technologies, today these sources contribute virtually nothing to the global energy mix.

The primary reasons for fossil fuels' dominance in today's energy marketplace are their abundance and apparent low cost. Our challenge going forward is not that fossil fuels are a limited resource, though that is true, but rather that their continued use is associated with tremendous hidden costs (see Chapters 3 and 14).

Skeptics argue that the recent trend of shifting from coal to natural gas (methane), facilitated by a new technique called hydraulic fracturing, which enables extraction of gas and oil from shale deposits, will alleviate climate disruption. This is a dangerous line

---

*As a review, a watt is a unit for the rate of energy use, which is a unit of energy (a joule) per unit of time (a second). So a watt is not a measure of energy itself, but rather a measure of how quickly energy is used.

of reasoning. While it's true that natural gas has a lower carbon intensity (lower $CO_2$ emissions) than coal, and it certainly releases less particulate, sulfur dioxide, and heavy metal pollution, in order to mitigate severe climate disruption, the global energy mix must have an average carbon intensity lower than that of natural gas (which is the fossil fuel with the lowest carbon intensity). You can't achieve that by switching to natural gas. Making matters even more troublesome, current natural-gas operations end up leaking large amounts of methane into the atmosphere, and methane is a much more powerful greenhouse gas than carbon dioxide. Recent studies suggest that the shift from coal to natural gas might actually be making things *worse* from a global-warming perspective!

The only true solution to climate disruption is to reduce our use of fossil fuels significantly. But if we want to minimize fossil-fuel use in the future and still allow for 30 TW of energy use by 2050, where are we to turn?

Nuclear energy could, in principle, be scaled up dramatically to meet a large fraction of this demand. However, nuclear, too, has some significant hidden costs. Foremost among these are (1) nuclear weapons proliferation and nuclear materials' use for terrorism (currently these are prevention and risk-management costs, but unfortunately they may shift to far more serious costs down the line), (2) long-term management of high-level radioactive waste, and (3) nuclear accidents.* While these accidents are very rare, their impact and cost are colossal. The Chernobyl accident in 1986 resulted in damage so severe that the region is still unusable. More recently, the Fukushima Daiichi accident in 2011 is projected to take a century to clean up and has an anticipated cost of a good fraction of one trillion dollars (with a "t"). That's big enough that maybe it shouldn't really be called "hidden" after all! Clearly, nuclear energy is here to stay, but society has to decide what the scale of its participation in the future energy mix should really be, considering all the factors.

---

*Oh, and (4) those three-eyed fish in *The Simpsons*.

Once you minimize fossil and nuclear energy, the only remaining energy options are the renewable sources. As we said, today these represent a small fraction of the energy mix, but what is their potential—their feasible potential, that is, not some romanticized view of their potential?

While you often hear about the various renewable sources spoken about interchangeably in the general media, not all renewable sources were created equal. Hydroelectric power, by far the most utilized renewable energy source today, is limited to at most maybe 2 TW globally because there are only so many rivers that can be effectively dammed or used in so-called run-of-the-river power plants (these latter ones don't require a dam that stores water). Wind represents a larger potential resource than hydropower (and represents the majority of new power generation in the United States in recent years), with a maximum feasible supply in the neighborhood of 6 TW. There is lots more energy than this contained in wind currents worldwide, but the wind speed has to be fast enough for a long enough time to be worth collecting with turbines, and there are only so many places on Earth where that is the case. Geothermal and ocean (tidal, wave, and thermal or salinity-gradient) energy can only contribute at the margins, with feasible supplies of around 1 TW and 0.2 TW, respectively. Geothermal energy is based on the heat from deep inside the Earth, which is a combination of heat left over from the formation of the planet billions of years ago and heating from radioactive decay of minerals. To access geothermal energy, you have to drill down far enough to reach those high temperatures, and that's only feasible where the Earth's crust is thin enough (like in volcanic regions), which doesn't include too many places. Biomass energy may play a large role in liquid-fuel production, but it will be in constant conflict with limited resources of land, water, and fertilizer already needed for food production. These are all wonderful sources of energy that we should take advantage of wherever possible, but even if fully utilized, together these sources wouldn't even represent a third of the projected global demand in 2050.

The one remaining renewable source is solar energy. Actually, most of the energy sources we just discussed are solar energy, too, just in disguise. Hydropower comes from the sun's evaporating water, which then falls as precipitation and runs down rivers, some of which can be dammed and used for electricity production. Wind happens when the sun heats up the air, causing it to rise, which in turn is replaced by cooler air.* Wave energy is driven by the wind (which is driven by the sun). Biomass energy comes from plants using sunlight to perform photosynthesis. Even fossil fuels are really solar energy, though they represent solar energy that was stored many millions of years ago in plants and other life-forms alive back then. Transforming solar energy into these other forms inevitably results in losses of useful energy (this gets back to that second law of thermodynamics that we discussed in Chapter 13), which suggests that capturing the solar energy directly should be more efficient. But how big is the feasible potential for solar?

You can do a simple back-of-the-envelope calculation to estimate the answer to this question. Start with the amount of sunlight hitting the surface of the Earth, which is a whopping 96,000 TW. (Recall that we currently use energy at a rate of about 18 TW worldwide.) Impressive as that number is, of course we can't use anywhere near all of that energy. For example, covering the oceans with solar panels is crazy. (This hasn't stopped entrepreneurs from trying to do it, however.) If you restrict yourself to solar energy hitting land, the resource drops down to about 28,000 TW. Still huge, but also still not feasible. We obviously need lots of that land for other things, like growing food. An aggressive but reasonable amount of land to use for solar energy collection might be 2 percent of the land. (This is approximately the proportion of land in the United States covered by roads.) Now we're down to 560 TW. Still huge, but still not feasible.

---

*The majority of the sun's heating of the air occurs at lower latitudes. Thermal gradients (variations in air temperature at different latitudes/locations) are thereby created that also drive wind.

An additional challenge is that, just as you lose useful energy when you convert sunlight to wind or biomass or hydropower, you also lose useful energy when you convert sunlight directly into electricity, heat, or chemical fuel. Yup, thermodynamics again. However, the losses in these processes aren't as big as those involved in transforming solar to other renewable energy sources.

A reasonable estimate for the average efficiency of converting sunlight to electricity would be about 12 percent. This is a pretty conservative estimate. There are commercially available solar panels that convert sunlight to electricity with nearly twice this efficiency. But if we go with this conservative number, our feasible supply drops down to 67 TW. Kinda feels depressing when we started with 96,000 TW, right? But that measly little 67 TW that could feasibly be extracted from solar energy is more than twice the total projected global demand for the year 2050!

The point is not that we should aim to get all of our energy directly from the sun, but that renewable energy sources are truly capable of meeting all of our energy needs, and that solar will need to be the biggest player in the future.

So why aren't we tapping into these low-carbon resources more than we are today?

Brad argues that renewables are just too expensive. His argument is predicated on ignoring the hidden costs associated with

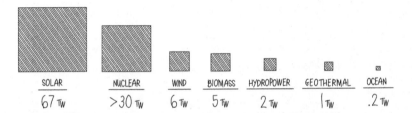

FEASIBLE LIMITS (APPROXIMATE) OF ENERGY SOURCES

| SOLAR | NUCLEAR | WIND | BIOMASS | HYDROPOWER | GEOTHERMAL | OCEAN |
|-------|---------|------|---------|------------|------------|-------|
| 67 TW | >30 TW | 6 TW | 5 TW | 2 TW | 1 TW | .2 TW |

**Feasible limits (approximate) of various energy sources**

traditional energy sources. If you account for those hidden costs, there is little doubt that drawing on renewable energy sources is considerably cheaper than relying on fossil fuels.* This plain fact busts Brad's case. But let's play Brad's game and examine the apparent costs anyway.

The first thing we need to do is figure out how to calculate the costs of different energy sources in a way that is comparable—apples to apples, if you will. Current electricity prices are the easy part. We know exactly what we're paying for electricity from the grid (in the United States, that's primarily coal, natural gas, and nuclear energy) because it's right there on our electric bill. That price varies widely from one location to another. For example, in Seattle, Washington, retail electricity will run you about seven cents per kilowatt-hour (kWh), whereas in New York City it's nearly eighteen cents/kWh. If you want to compete with traditional energy sources, you need to have a cost that is at or below this number where you live. The term used to describe this competitive price is "grid parity." Achieving grid parity is much easier in some places than in others.

Figuring out what a new energy source will cost isn't as easy. The best way to calculate the cost of an energy source is to determine what's known as the levelized cost of energy (LCOE). You can view the LCOE as the price at which electricity must be generated from a specific source to break even over the lifetime of the project. It will be the lifetime cost for that energy project divided by the energy it produces in that lifetime.

For example, a solar-panel system will probably last at least thirty years. You can estimate how much electricity you'll get from the panels over that time by projecting the amount of sunshine in the location where they'll be installed and by knowing their efficiency at converting sunlight to electric power. You also need to

*Even with the hidden costs exposed, humans tend to be poor individual decision makers when it comes to long-term risks. This is one reason why so many people smoke cigarettes, even knowing they are putting themselves at increased risk for heart disease, lung cancer, and other serious health problems.

know how much the performance will diminish over time. Most of the cost for a solar-energy system is up front: You have to buy the panels and their mounting hardware, you have to get something called an inverter that shifts the DC electricity generated by the panels into AC electricity that you use in your home or business, and you have to pay folks to install everything.* There will also be some additional costs over the twenty-five- to thirty-year lifetime, like maintenance, repair, and occasional cleaning. Putting all this information together, you can project the LCOE for the system, and the number you get will be in units of ¢/kWh. There's your apples to compare to the grid apples.

Brad pipes up, "Wait just a second! There are government subsidies for those solar energy systems, so the comparison isn't fair. You're cheating . . . it's really apples to oranges."

Yes, Brad, many governments offer tax credits or other incentives to encourage homeowners or business owners to install renewable energy systems. Subsidies are an important topic that we'll get to in a bit, but first let's be fully transparent here and look at both the subsidized and the unsubsidized costs of something like a solar-energy system side by side with grid-electricity costs.

Costs for renewable energy systems like solar and wind have historically been high (like Brad claims), but they have been dropping pretty steadily for years. Grid prices, in contrast, have been rising steadily. Both of these trends are expected to continue for a long time, which suggests that grid parity is inevitable.

The point in time when grid parity is reached will depend on where you live. This is true in part because, as we mentioned earlier, electricity costs much more in some places than in others. But you

---

*Because the costs are front-loaded like this, many families cannot afford to install a solar-energy system on their home even if they understand the long-term savings. Fortunately, there are myriad business models out there that can defray these up-front costs, such as by spreading them out over the lifetime of the system or by creating a system that allows the homeowner to effectively buy only the electricity produced by the equipment without having to own the equipment itself.

# INEVITABLE GRID PARITY

──── UNSUBSIDIZED SOLAR
──── SUBSIDIZED SOLAR
----- RETAIL ELECTRICITY
○ GRID PARITY

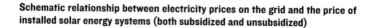

COST

TIME

**Schematic relationship between electricity prices on the grid and the price of installed solar energy systems (both subsidized and unsubsidized)**

also have to consider the potential costs of the renewable energy sources. So, sticking with our solar-energy example, some regions are much sunnier than others, which means that you'll get more energy from the same panels in sunnier places, which will make the LCOE for those systems much lower. In other words, if you live somewhere that is sunny and has high electricity prices, like California, you're already at *unsubsidized* grid parity today. (That probably explains why solar panels are popping up like dandelions in California these days.) Those of us living in places with somewhat less sunshine and lower retail electricity prices aren't there yet. In some of these areas, subsidized grid parity has been achieved with solar energy systems, while in others it hasn't happened—yet.

Of course, it hasn't happened yet if you're comparing the LCOE to the retail electricity price today. Assuming that electricity prices continue to rise in coming years (no one seems to doubt this fact), and remembering that your system will last for decades, you will almost definitely save a bunch of money in the long term by switching to renewable sources now, even if you lose some in the first few years.

There are additional economic reasons for why renewables are becoming increasingly more affordable. Some utility companies have started offering "time of use" pricing. This means that the price you pay for a kWh of electricity varies over the course of a day, depending on the overall demand at the moment. For example, if it's an oppressively hot and humid afternoon and lots of people switch on their air conditioners, the price for electricity will go up. In fact, it is almost always sunny afternoons when prices are highest, and this is also when solar panels are producing the most electricity. That means that solar can provide more economic value by reducing demand and providing extra electricity when it's most needed. This phenomenon moves us closer to achieving grid parity.

OK, so we've explained that Brad is wrong when he says that unsubsidized renewable energy is too expensive—at least in certain regions (for now). Renewables do wonderfully in the apples-to-oranges (subsidized) comparison, but they even win in some apples-to-apples (unsubsidized) comparisons today, and they'll win more and more as time passes.

Skeptics routinely criticize these government subsidies for alternative energy, claiming that they interfere with the free market and unfairly position renewables as more favorable options. Let's take a closer look at energy subsidies.

In the United States, federal subsidies to foster energy innovation are not something newly invented for renewables.* Any time there has been a transition from one energy source to another, government subsidies have played a major role. This started with land grants for timber and coal in the nineteenth century, moved to tax expenditures for oil and gas in the early twentieth century, then to support for nuclear in the mid-twentieth century, to biofuel (ethanol) subsidies in the late twentieth century, and finally those solar and wind tax benefits we have today. So this is standard operating procedure for the federal government. But are today's subsidies outliers in terms of their size?

Yes, but not in the way Brad would have you believe. Adjusting for inflation and looking at the first fifteen years of each subsidy program, we find that nuclear received 3.3 billion dollars per year, oil and gas 1.8 billion dollars, and renewables only 0.37 billion dollars. So those subsidies for renewables are actually far smaller than the generous amounts historically afforded to other new energy sources. Moreover, the social cost of carbon that we discussed in Chapter 14 is essentially another gigantic subsidy for fossil fuels—one that is paid by everyone on the planet. So Brad's apples-to-oranges argument is really more of an apples-to-giant-mutant-grapefruits situation, with renewables getting the short straw.

We've busted the renewables-are-too-expensive myth, but skeptics have another major criticism for alternative energy, namely that it can't provide baseload power. What they mean is that energy sources like solar and wind are variable—the sun doesn't shine all the time and the wind doesn't blow all the time. Baseload power plants run constantly and provide stable, uniform power to the grid. The skeptic argument says that, once renewables are a larger fraction of the energy mix, their variability will cause widespread blackouts and the like.

---

*DBL Investors, a venture capital firm that focuses on both financial success and social impact, released a report in 2011 entitled *What Would Jefferson Do?* that provides a quantitative comparison of historical energy subsidies in the United States.

There are two answers to this critique. The first is that some renewable energy sources actually can provide baseload power. Wind compressed air energy storage (CAES) plants, some of which have been in operation since the 1970s, work by storing excess energy that's generated when the wind is strong as compressed air in underground caverns. Later, when the energy is needed, the compressed air is discharged, which turns turbines and generates electricity. Solar thermal plants focus the sun's rays on a fluid, which gets really hot (think magnifying glass and an ant on the sidewalk). That heat can be used immediately to boil water, make steam, and turn turbines, just like in a traditional coal or nuclear power plant, or the heat can be stored for many hours and used to generate steam at a later time. Hydroelectric, geothermal, and biomass plants are other examples of renewables with baseload power generation capability.

The second answer to variability is that there are many strategies available to accommodate it. We'll briefly go over five: forecasting, interconnection, scheduling loads, oversizing, and storage.

**FORECASTING:** The grid has not had to contend much with supply variability in the past, but it has always had to handle demand variability—like when folks switch on those air conditioners all around the same time. The problem is that most baseload power plants require hours to ramp up from being turned off, which is clearly too slow to deal with short-term demand variability. In order to avoid power disruptions, grid operators have integrated something called operating reserve. There are two flavors of operational reserve: spinning reserve and supplemental reserve. The former usually involves extra capacity in the power plants that are already powering the grid, and the latter involves power plants that can be fired up rapidly. So spinning reserve involves power plants that are running basically all the time and throwing away some of their energy, but that can be called upon on short notice when demand spikes (turn it up to 11, Spinal Tap!).

Adding variability to the generation side through large amounts of solar and wind power will exacerbate the problem, and increasing backup power options, which typically use natural gas as a fuel, defeats the purpose of switching to renewables in the first place.* Improved forecasting of cloud cover (for solar) and wind speeds (for wind) could help substantially reduce the need for wasteful operating reserve practices.† If you know ahead of time that some clouds are going to pass over your big solar array, you can fire up your operating reserve just when you need it rather than having it running all the time.

**INTERCONNECTION:** Our electric grid is a marvel of humankind's engineering talent, but it is a remarkably outdated technology. You've probably heard talk of a future "smart grid." This term can mean many different things, depending on the context. One meaning is a more interconnected grid. How would this help with variability in power generation from renewable energy sources? Well, maybe the wind isn't blowing near St. Louis at the moment, but it might be in Chicago (it is the Windy City after all . . .). So you could move power from one region that's generating more than it needs to another that's coming up short. Today, the grid is not well equipped to move power on a large scale and over large distances like this, but there's no reason why it couldn't do this in the future.

---

*An additional myth cited by skeptics is that cycling fossil-fuel power plants to accommodate variations in renewable energy output will result in an overall *increase* in emissions. (The argument is based on the fact that power plants emit more when they are powering up than when they are running at full power.) A recent study conducted by the National Renewable Energy Laboratory, however, found this claim to be utterly baseless. Carbon emissions induced by the more frequent cycling are found to be less than 0.2 percent—a negligible increase dwarfed by the overall reduction of more than 30 percent due to the replacement of fossil fuels with renewables.

†Improved forecasting of this nature is an active and promising research area at many institutions, including our home base of Argonne National Laboratory.

**SCHEDULING LOADS:** Some things need to turn on the moment you flip the switch. You need the ceiling light and your computer to respond immediately. On the other hand, some things can wait. For example, if you commute in an electric car and plug it in when you get home, it probably needs about four hours to charge up, but you likely won't be driving it until the next morning. A smart grid would monitor the available supply of electricity and draw power for your car batteries only when there is enough available; that is, when the renewable sources are generating more than is needed. Other examples of loads that could be scheduled are clothes washers and dryers, water heaters, and dishwashers.

**OVERSIZING:** Variability in power production from renewables is a concern only if the power generated is sometimes lower than the overall power demand. We will need 30 TW of energy in 2050, but what's to stop us from building an energy infrastructure with more than that? If we had capacity for, say, 32 TW, then when there is a drop in variable energy supplies because of a particularly cloudy or calm day, the extra capacity would be available to fill in the gap.

**STORAGE:** Probably the most effective way to deal with variable energy sources is to store excess energy when it's available and draw on that storage when needed. Techniques used in CAES and solar-thermal plants are good examples of how we might implement this approach, but they aren't the only ones. Another way to store energy is to pump water up to a reservoir at higher elevation. It can sit there for an indefinitely long period, until it's needed, at which point it can be released to flow back down through turbines to generate power. Or, you could store electricity in batteries. There could be large banks of batteries devoted to this purpose connected to the grid (you can purchase such a system for your house today to completely remove yourself from the grid if you have on-site power generation such as from solar panels). As more people drive plug-in vehicles, these, too, can serve as a massive electrical-energy-storage reservoir, since some fraction of them will always be plugged into outlets on the grid. You can also use excess power to perform chemical reactions, thereby generating fuels that could be burned at a later time. Using electricity (from wind or solar) to split water into hydrogen and oxygen gases by electrolysis does this. When burned, hydrogen gives you heat and—as a "waste" product— water. Sure beats toxic heavy metals, particulates, and carbon dioxide!

Together, these strategies could help renewables to penetrate the energy market on a scale that would truly make a difference in mitigating climate disruption. Skeptics make all sorts of arguments about why we should just stick with the old energy sources we've got today, but they're wrong. Fossil fuels are unsustainable and causing tremendous damage to both the planet and our society. Renewables aren't too small. They aren't too expensive. They aren't too variable. They are (a huge part of) the answer.

# Epilogue

## OUTLOOK

**WHAT CAN BRING** together a world-renowned scientist, a Saudi oil minister, and Dr. Seuss? Climate change, of course, as we'll explain.

Skeptics seem hell-bent on proving that the climate isn't changing and that, even if it is, humans have nothing to do with it. As we've discussed in this book, the science says otherwise. Human activities are changing our planet's climate. Atmospheric concentrations of greenhouse gases like carbon dioxide have risen dramatically starting with the Industrial Revolution, and the burning of fossil fuels is strongly correlated with this increase. Those greenhouse gases, which will remain in the atmosphere for hundreds if not thousands of years, are the primary reason that the Earth has warmed by about 0.8 degrees Celsius over the past century. No known natural phenomenon can provide an explanation for this warming. Particularly alarming is the fact that many aspects of climate change are accelerating as a result of both increasing carbon dioxide emissions and positive feedback cycles. We ain't seen nothin' yet.

Global warming means far more than just higher temperatures; its consequences reach essentially every life-form on the planet. We're already experiencing increased extremes of heat, coastal flooding, and precipitation, and all of these will progressively get worse as warming proceeds. Diseases will spread to new regions, already-limited water resources will be further strained, infrastructure will suffer greater damage, and crop productivity will plummet. Ocean acidification, the sinister cousin of global warming, will wreak havoc on biodiversity in the seas. Biodiversity on land, too, will suffer as habitats are transformed in response to climate disruption. The pace of these changes will be far too fast to allow many species to adapt.

The scientific community is certain of all these things. There are uncertainties with regard to exactly which impacts will be experienced at any given location on Earth, and at what time, but there is no uncertainty that these things will happen.

We can all agree, whichever church, mosque, temple, or synagogue we attend, or if we don't attend at all, and whichever party we vote for—that we want our children's generation to be better off than our own. Every day that we continue business as usual with fossil-fuel emissions, we are putting this future at greater risk. We are in this together. Climate disruption is probably the greatest challenge we've faced as a society, and if we don't start dealing with it aggressively as a society, it will get the best of us. Climate disruption doesn't care what political party you belong to or even whether you believe that the climate is changing at all. It will hit you if you're in Beverly Hills or the shantytowns of Karachi. And it's relentless.

In the face of this bleak future, it's tempting to throw in the towel. It feels like we've created such a mess for ourselves that we won't be able to dig our way out. Daniel Nocera, an American scientist whose research is aimed at creating an artificial leaf—a system that, like photosynthesis, can use the energy from the sun to make chemical fuels—visited Argonne National Laboratory (our place of work) outside Chicago, Illinois, in 2010. He spoke to this daunting challenge we face as a society and said, "The obligation as a scientist is to go down swinging." That can be broadened. It's our obligation as humans to step into the ring.

What can we do? Waiting until fossil fuels peter out on their own isn't going to cut it. There are plenty of them buried down there, but as Sheikh Ahmed Zaki Yamani, a former Saudi Arabian oil minister, famously said in 1973, "The Stone Age didn't end because we ran out of stones." The Stone Age ended because better technologies came along. Just as bronze replaced stone, alternative energy technologies can replace fossil fuels.

Shifting energy regimes is a monumental challenge, but it's one that we can—we must—overcome. The energy revolution has begun. In 2013, renewable energy sources provided 14 percent of US electrical generation. That's great progress, but we have a long, long way to go. History has proven that the market won't get us there on its own—at least not quickly enough. The public—all of us—need to engage with this challenge. As Dr. Seuss' Once-ler in the 1971 classic *The Lorax* reminds us, we have to stand up for what's important to us, because if we don't, nothing will ever change.

It goes beyond turning off the lights when you leave a room, or carpooling to work. Skeptics standing in the way of solutions are asking us to stick our heads in the sand. We've laid out the facts that reveal these skeptic arguments as smoke and mirrors; they are distractions from the problem that grows more pressing with each passing day. We can fix the mess we've made of our climate, and doing so is the only way to give our children's and our children's children's generations even a shot at having a bright future.

# Acknowledgments

CLIMATE SCIENCE IS a broad, deep, and continually evolving field of research, and the skeptic myths surrounding it are themselves fantastically diverse and incessantly changing. A book of this nature would never have been possible without the generous input and feedback from a collection of friends and colleagues to make sure that the climate science was up to date, accurate, and digestible enough for the general public. Climate data, climate-change phenomena, and impacts of a changing climate are all complex in their own ways and quite a challenge to convey. Sarah Sisterson, our talented illustrator, played a pivotal role in achieving this goal with her stylish depictions of everything from ice-core data to economics.

Researching and writing this book on nights and weekends would never have been possible without the support and patience of our families. We dedicated this book to our children, Isaac Darling and Nathaniel, Sarah, and Rachel Sisterson, whose generation will inherit the climate challenge. Our agent, Peter Tallack at The Science Factory, took on this manuscript when most of his peers were scared off by the prospect of a climate-change book, and he was instrumental in distilling what made it unique. We were also fortunate to have Nicholas Cizek as our editor at The Experiment; he tolerated our penchant for rigor and nuance at each step in the process.

Importantly, we would also like to thank the myriad scientists who toil thanklessly every day to better understand our climate and our impact upon it. It is only through their relentless efforts that this issue has gained the public attention it so desperately needs.

# Index